David Hall was Fred Dibnah's TV producer for many years, as well as a friend of the family. He made all of Fred's most recent television series and is also the producer of *Fred Dibnah's World of Steam, Steel and Stone*.

Also by David Hall

FRED DIBNAH'S INDUSTRIAL AGE
FRED DIBNAH'S MAGNIFICENT MONUMENTS
FRED DIBNAH'S AGE OF STEAM
(all with Fred Dibnah)

FRED

The definitive biography of Fred Dibnah

DAVID HALL

CORGI BOOKS

FRED
A CORGI BOOK : 9780552154888

Originally published in Great Britain by Bantam Press,
a division of Transworld Publishers

PRINTING HISTORY
Bantam Press edition published 2006
Corgi edition published 2007

9 10 8

Set in 12/15pt Times New Roman by
Falcon Oast Graphic Art Ltd.

Corgi Books are published by Transworld Publishers,
61–63 Uxbridge Road, London W5 5SA,
a division of The Random House Group Ltd.

Addresses for Random House Group Ltd companies outside the UK
can be found at: www.randomhouse.co.uk
The Random House Group Ltd Reg. No. 954009.

Printed in Great Britain by
Cox & Wyman Ltd, Reading, Berkshire.

The Random House Group Limited supports The Forest Stewardship
Council (FSC), the leading international forest certification organisation.
All our titles that are printed on Greenpeace approved FSC certified paper
carry the FSC logo. Our paper procurement policy can be found at:
www.rbooks.co.uk/environment.

To my wife, Fran,
with love and thanks
for her support

In memory of
a friend and
much-loved steeplejack

Contents

Acknowledgements

It was one of the major privileges of my life to know Fred Dibnah and to have made so many television programmes with him. Writing this book and making the television series that complements it has been in the main a highly enjoyable process, bringing back many happy times with Fred. My thanks are due first and foremost to Fred not just for this and for bringing pleasure to the lives of millions of viewers but also for telling me so many times that he wanted me to write his biography and providing me with so much material for it. If it hadn't been for Fred's wish, this book would not have been written.

I am particularly indebted to Fred's widow, Sheila, for all her help and encouragement, for the interviews she has given and for verifying important details. I wish her every success with the book she is planning on life with Fred. Special thanks are also due to Fred's brother, Graham, for filling in invaluable detail about Fred's early years and their family background.

I would also like to thank the many friends and colleagues of Fred who have contributed to this book and to the tribute programmes I have made for the BBC. The list would be too long if I were to name all of them but I would like to make special mention of Neil Carney, Eddy Chatwood, Alan Crompton, Jimmy Crooks, Alan Davies, John Gourley, Ian Howard, Alan McEwen, Tom McFarlane, Jack Meaker, Alf Molyneux, Derek Roscoe, Colin Shaw, Michael Webber, Valerie Taylor and Barbara Yates. Many of the above have provided pictures for the book from their own collections, for which I thank them. In addition I'm particularly grateful to Paul Donoghue for providing many of the photographs from Fred's early years.

Special thanks to my daughter, Kathryn, for all the backup and research she has done for this book and the television programmes, for setting up and conducting most of the interviews, organizing all the transcripts, finding archive photographs and above all for handling with tact and diplomacy the sensitive task of liaising with members of Fred's family.

Thanks also to Chris Matthews for his work on editing the transcripts, providing invaluable historical background and collating research notes, and to Aisha Mir for additional research; to Mark Lesbirel of RDF Rights for prompting me to write the book and for setting up the publishing deal; and to the team at Transworld, especially Doug Young for commissioning the book, Simon Thorogood for scouring the text for ambiguities, inconsistencies, repeats and omissions and Deborah Adams for getting my text into presentable form.

Without his television programmes and the fame that came with them there would have been no book.

ACKNOWLEDGEMENTS

Television production is a team effort. Nobody was more aware of this than Fred, and I know he would want me to thank the people who made him what he was – Don Haworth and his team of Jean Thompson, cameraman Arthur Smith, sound recordist Jack Wilson, film editors Peter Gibbs and Roy Newton, and Martin Lightening, the cameraman who went up the chimneys with him to get those spectacular shots: and my team of assistant producers Bob Sandy and Jon Doyle, cameramen Andrew Hartley, Rob Taylor and Mark Hamilton, sound recordists Nigel Chatters and Alan Parry, film editor Steve Parry and dubbing mixer Chris Rayson, who eased Fred through the difficulties of recording his commentaries.

Above all, thanks to the BBC for bringing so many of Fred's programmes to our screens for over twenty-five years. For the more recent series I would like to make special mention of Richard Klein and Mike Greenwood and thank them for their unfailing support for the programmes within the BBC. Without them we wouldn't have seen anything like as much of Fred on our screens as we have done.

Finally I would like to thank my wife Fran for all her behind-the-scenes work and for her support over the years, especially over the last nine months while I have been writing this book and making *Fred Dibnah's World of Steam, Steel and Stone*. I should now be able to help her with some of the work on our new house, which has recently all been left to her so that I could meet my deadlines.

Introduction

Fred Dibnah – steeplejack, steam enthusiast, television presenter – was many things to many people but you'd be hard pushed to find anyone who didn't admire his commitment to Britain's industrial heritage. He longed to tell everybody who cared to listen how things were built and how they worked. Over a period of seven years, from early 1998 until his untimely death from cancer in November 2004, I made more than forty programmes for the BBC with Fred. During this time he became much more than a presenter of our programmes; he was a good friend whom I would see throughout the year, not just in the months when we were filming.

Fred and I hit it off from the first time we met. We came from the same working-class background in the north-west and grew up when places like Bolton and Manchester, where I was from, were still alive with manufacturing industry. I never really understood how Fred's engines worked but we shared the same passion for telling the stories of the engineers whose skill and inventiveness made them possible, and recounting our history from the point of view of the builders and engineers and the ordinary working men who helped to make it. For seven years Fred and I spent every summer travelling the length and breadth of Britain, visiting magnificent castles and cathedrals, climbing to the top of great bridges like the Forth and the Humber, searching out steam-powered mill engines and pit winding engines, tunnelling beneath the earth in coal mines and tracing the routes of long-forgotten canals. For Fred it was all a holiday paid for by the BBC, taking him to all these wonderful places that he wanted to go to. It was in that spirit that we made the programmes, so that his enthusiasm and sense of fun were never diminished. But filming with Fred wasn't without its dangers. Climbing to the top of the dome of St Paul's with him wasn't a problem, but driving was. As we sped along the motorways I'd often struggle to keep going in a straight line as Fred would dig my arm to point out a chimney. Wherever we went, we'd always go for a drink together at the end of the day and I've lost count of the number of lock-ins we had as the pints were lined up on the bar while Fred chatted with his many fans.

Fred was already a well-established television personality long before I met him. Through the wonderful films of Don Haworth that chronicled Fred's life as steeplejack and steam enthusiast and his marital ups and downs, he had become what many people described as a 'national treasure'. But by 1997, when I first met him, his television career, along with his steeplejacking, was in decline.

There were not many chimneys left to restore or to demolish and, as the work began to dry up, the broadcasters started to feel that they had gone as far as they could with Fred Dibnah.

But the one thing television needs more than anything else is presenters with passion, enthusiasm and personality and, when it came to industrial history and engineering, it was clear that Fred Dibnah had it all. So, although up to this time he'd only ever been the subject of documentaries, I put a proposal to the BBC for an industrial history series with Fred as presenter. *Fred Dibnah's Industrial Age* was commissioned in 1998 and from that moment on Fred had a new career. But Fred was unique – unlike any other television presenter before or since. As David Robson wrote in the *Sunday Telegraph*, 'His genius lies in cutting through the stiltedness of so much documentary television and speaking to the viewer in simple colloquial English. There is a whole world separating the glitz of the television studios from everyday life. Very few programmes have bridged that gap, have been able to illuminate ordinary lives in the way that Fred Dibnah does.'

Fred was from Bolton, Lancashire, and he was proud of it. And it was Bolton that provided his inspiration. Growing up there in the 1940s and 1950s, he fell in love with the decaying industrial landscape around him and he developed the interests that were to stay with him for the rest of his life – steeplejacking and steam engines. He saw beauty in the mills and the chimneys and the engine houses, and especially in the great steam engines inside them. It was in these early years that Fred first became interested in climbing chimneys.

Steeplejacking wasn't his first job. Fred served his apprenticeship as a joiner although, all the time, despite parental disapproval, he pursued his goal of becoming a steeplejack. Over many years he studied how steeplejacks erected their ladders and scaffolding, noted the faults in different systems and worked out his own methods.

But another passion was to play just as important a part in his life. When he was a lad, the engine sheds were just down the road from the terraced house he lived in and he developed a dangerous addiction – steam. It was his memories of those engine sheds that got him interested in steam preservation and led to his buying a steamroller and restoring it. Fred didn't like modern methods and new-fangled tools, so he decided to equip himself with a workshop powered by steam. He added to it over the years and created something unique.

Until 1979, Fred was only a steeplejack who had an interest in steam engines. Then, while he was working on the clock tower of the town hall in Bolton, the local BBC TV news programme filmed a short item about him. Following this, Fred was approached by television producer Don Haworth with a view to making a half-hour film for a series about people with unusual occupations. After several months of filming, the finished article was an hour-long documentary, *Fred Dibnah, Steeplejack*, which won a BAFTA award.

From that time on, Fred has rarely been away from our screens. He became for millions of viewers the embodiment of old-fashioned common sense. Straight-talking and frank-speaking, Fred always gave you what it said on the tin and his forthright views on anything and everything turned him into something of a folk hero. Through his early programmes Fred became best known for felling chimneys. But this was the job he liked least. His preference was always for restoration and he would point proudly to chimneys and church steeples that he had restored.

When he turned to presenting, his blunt, no-nonsense style and the direct, colloquial English he used made a welcome change from the majority of television presenters. The audience didn't only enjoy watching Fred, they trusted him because he was one of them. Few people on television have been better at talking to ordinary working people and putting them at their ease. But what we loved more than anything with Fred was his enthusiasm. He had a passion for anything to do with our industrial past and an infectious zest for the engineering, mechanical and craft skills of days gone by. His heroes were the great engineers of the Victorian Age such as Robert Stephenson and Isambard Kingdom Brunel. In all of his programmes Fred's approach was very hands-on: he lived every schoolboy's dream, travelling on the footplate and driving the engines. But there was always a sense of sadness alongside the passion. He felt keenly the loss of the sights and sounds of long-gone industries.

Steam and engineering weren't Fred's only loves. Because he'd worked with wood he could talk with authority about the skills of carpenters and woodcarvers when we filmed at great castles and palaces and country houses. He loved meeting men like Peter McCurdy, who built the timber frame for the reconstruction of Shakespeare's Globe Theatre. While admiring the finished products, he was just as interested in the way the job had been done and in working conditions and he was always able to put himself in the place of the workers.

Whatever the job, Fred would be prepared to have a go at it himself; and whatever he saw, he'd always got one like it at home or in his back garden. As a young man he had been to art school and the drawings he used to show how things were built were all his own. His practical demonstrations revealed exactly how things were done, even if some of them did end in disaster. And because he'd been there and done the job himself, Fred could speak with an authority that few academics could share, and bring his own unique insights to everything from a medieval siege to the generation of steam in a nuclear power station.

For me it was a real privilege to make programmes with Fred. One of British television's greatest characters, he was also a lovely man to work with and go for a pint with at the end of a day's shoot. And he was totally untouched by fame or celebrity. Wherever he went he always found time to chat to anybody, to sign an autograph or pose for a photo. Every one of his old friends said fame never changed him; he was still the same Fred that he had been twenty or thirty years ago when they first met him. There was never anything remotely showbiz about him. Right up to the end his mates were engineers and miners and steam men; he respected men who made things, who mined the coal or drove the engines that powered the wheels of industry. These were the men Fred wanted to spend his time with, not people from the world of television or show business.

The audience trusted him and believed what he said because he wasn't a posh television presenter just reading a script. He was a working man who spoke from practical experience when he talked about how things were made or built. Fred enriched the lives of those who worked with him and were his friends. There were a lot of laughs when Fred was around; and everything was always a bit more dangerous.

Much of his appeal came from his obvious love of life and his sense of humour. He always had a smile on his face and in all the time I worked with him I never heard him complain about anything. Fred was a great raconteur with a never-ending fund of stories, a larger-than-life character who could walk into a pub and hold an audience enthralled until the beer ran out.

Fred embodied everything that was wholesome from what many people perceive as a more wholesome era. He represented old working-class values of hard graft and pulling yourself up by your bootlaces – values that are largely missing from modern society. He was enthusiastic about a way of life that has virtually disappeared now and he made no apology for living in the past. For many people, he epitomized the affection they felt for the architecture and machinery of Britain's once-great manufacturing empire. With his flat cap and his working clothes he was a working man who typified the northern mill towns of the 1950s. If you went into the local pub fifty years ago, there was always a guy like Fred in the corner. He could have been a builder or a plumber, or he could have worked in a local foundry, an engineering works or a mill. Fred was the epitome of northern grit.

But Fred was also a complex individual, a man whose life and personality were full of contradictions. He spent much of his working life demolishing chimneys, but it was a job that filled him with sadness. He loved the chimneys and had a great appreciation for what to him had been the heroic efforts that had gone into building them. For Fred they were architectural treasures that should have been preserved.

He was tough and fearless and he revelled in a certain kind of macho image. Everything he liked, he used to say, was heavy, dirty or dangerous, but he was so sensitive that he would be moved to tears when he heard the news that a friend's engine had been destroyed in a fire.

He could be stubborn, selfish and driven by his own obsessions, but he was the most non-confrontational man you could ever meet and he was easily manipulated into doing things that other people wanted him to do. For Fred it was often a case of taking the line of least resistance, being persuaded by the person with the loudest voice, and saying the things he thought people around him wanted to hear.

Fred was a gentle, friendly character with no malice in him, but in his relationships with women he was something of a male chauvinist. He always claimed that he had been born out of his time, that he should have been born in the Victorian Age. For Fred a woman's place was in the home. His wife would be expected to get the cleaning done and have his tea on the table at a certain time. For all his three wives it led to problems and confrontations.

Although he was always surrounded by people, Fred was essentially a loner. As a boy he wouldn't join in with his brother and their mates to play football or Cowboys and Indians, but would go off on his own to look at chimneys or old engine houses. When he got older his mind was always on his next project, working on one of his engines or on his coal mine in the garden. But he was a loner who loved people. He enjoyed being the centre of attention, holding court as he told one of his stories. If you met and talked to Fred, he would treat you as if he had known you all his life. This meant that, as his fame spread, he became public property, and he always wore his heart on his sleeve. There was a certain naivety about him, which meant that he would open up his heart to any stranger who cared to listen. Through the medium of television those strangers ran into millions. Everywhere you went with him, people would come up to Fred and ask him about his wives. Nobody can have discussed their marriages and personal problems with more people than Fred.

It was these very contradictions and flaws in his character that made him so loved by so many people. But more than anything else, I believe his appeal came from the fact that he was the eternal schoolboy. With his collection of boys' toys he was the envy of every one of us who would have liked to spend our working lives playing with the things we played with as children. And we had a certain envy for that childlike self-centredness, even selfishness, that kept him focused on the things he wanted to do to the exclusion of all else. He was the lovable child who would never come indoors when his mother said it was time for tea. He was lucky, we felt, to be his own boss; to be able to do his own thing without worrying too much about what anybody else wanted; to be free to say the things he wanted to say without being too concerned about what anybody else was thinking; and to make a living out of doing what he was passionate about.

Fred loved having film and television cameras around him and the camera loved Fred. It was drawn to him as if to a magnet and Fred was always centre frame. He gave the distinct impression that he would have been happy to have a camera recording his every move, and I lost count of the number of times I had to turn down his requests to go to his house with a crew to film an important stage in the building of his traction engine or in digging his coal mine.

Fred was a joy for any producer to work with. You can't manufacture the passion and enthusiasm that he brought to the programmes we made together. He was a natural broadcaster, no different off camera from how he was on screen. And there were never any tantrums or awkward moments. He'd always go along with whatever he was asked to do and, although he enjoyed being the centre of attention, there was never any 'look at me, I'm a star' about him. There was only one complaint: 'All the best stuff ends up on the cutting room floor,' he would say. But Fred's idea of the best stuff was a detailed explanation of the workings of a mechanical lubricator, which was not the sort of thing that was suitable for a half-hour prime-time television series. So I promised Fred that I would make sure the unedited versions of his programmes would be made available to his fans on DVD.

Fred became the voice of a vast silent majority, a large slice of middle England whose views and aspirations are not reflected very well on modern television. He represented the values of a time when Britain still had manufacturing industries, when the concept of empire wasn't politically incorrect. He spoke with pride of a Britain that led the world – an idea that has become unfashionable today. He stood for the dignity of making things, the comradeship of the working man and the satisfaction of an honest day's work and a job well done; and he represented the values of generations of working-class people who believed in self-help and in education, the sort of people who always wanted their children to have a better life than them. These were the values that Fred felt we had lost and he thought that, as a nation, we were much the worse for it; and his views were shared by many, many people.

They are the values that Fred wanted to preserve and they will live on in his programmes and in all the things we filmed together. The most recent series, *Fred Dibnah's World of Steam, Steel and Stone*, is a fitting tribute and it will form a lasting record of the things he stood for. Making it and writing the biography he always said he would like me to write have brought back many happy memories of my time with him, and his distinctive tones still ring round our offices every day as we continue to edit some of his programmes.

Fred was Britain's unlikeliest superstar. As his death was announced on BBC News 24, presenter Peter Sissons remarked, 'They don't make them like that any more.' He was one of a kind and now he has gone we will never see anybody quite like Fred Dibnah again. As I worked with him and watched him do a drawing, make a model or come up with a unique solution to a mechanical problem, I often used to wonder, somewhat fancifully, whether Leonardo da Vinci had been a bit like Fred. Leonardo was an artist, an inventor and an engineer; a man with passion and energy who, just like Fred, didn't mind rolling his sleeves up and getting his hands dirty. History paints a picture of the hero of the Renaissance as someone associated with high culture, but perhaps Leonardo was more like Fred: a clever, artistic, inventive, highly skilled working man who was able to achieve great things. Perhaps Fred was not just Britain's unlikeliest superstar; maybe he was an even more unlikely Renaissance Man.

Part One
The Working Man

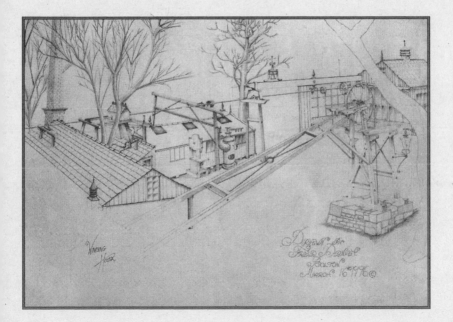

1

The Industrial Landscape

Fred was born in Bolton, Lancashire, in 1938 and his early years were spent in what he always described as a '*Coronation Street*-style house' next to the railway sidings. His father, Frank, worked at a bleach works and his mother, Betsy, was also employed there before she got a job as a charlady at the gasworks. They both had hard lives and never had much money when Fred was young, but it was a happy home.

Bolton was built on the cotton trade and when Fred was born the scale of the industry in the town was massive. He grew up surrounded by cotton mills and the great steam engines that provided the power for them. It was a sight that was to stay with him for ever, and when I started filming *Fred Dibnah's Industrial Age* he recalled it vividly. 'One of the first sights I remember was the factory chimneys that surrounded the town, especially all those near Bolton Wanderers' old football ground at

Burnden Park. When the sun was setting in the west, you could see them silhouetted against the evening sky above the area where I used to play as a kid. Then at night, when you looked in the dark at this great big five-storey spinning mill when the night shift was on, it was an unbelievable sight. All the lights would be on and you could see literally hundreds of yards of two-inch shafting with the chromium plate all shining in the dark.'

As a small boy he used to ride his bicycle along the Manchester–Bolton–Bury Canal with his father and to Fred the world around that towpath was an exciting place to be. It was a landscape of great, gaunt spinning mills and bleach works standing by the side of dirty old canals and railway tracks. Steam hissed from the engines and the smoke from the mill chimneys filled the air – dozens of them belching out thick, black clouds day and night, so that a grey canopy hung permanently over the cobbled streets and terraced houses. The chimneys towered above everything like tall grass in a meadow. They had to be high not just to carry the smoke and noxious fumes up into the atmosphere but also to create a draught for the boilers in the mills and factories down below.

There were so many chimneys that around seven tons of soot would fall on each square mile of Bolton every year. Thick smogs, or pea soupers as they were known, would envelop the town for days on end and it wasn't a very healthy place to live. If you went up on to the moors around Belmont to the north of the town, you couldn't see the place because it was completely covered in a thick greenish-yellow blanket. November was the worst time, always bringing the thickest smogs. But sometimes on a summer's day you could watch the shadows of the mill chimneys from up on the moors as they moved slowly, like the hands of a clock, over the streets, terraces and houses. The smog could get so bad that one of Fred's friends, Alan Crompton, remembers going to the cinema with his mum and dad, and when they came out they got lost and couldn't find their way home. They couldn't see the street name four or five feet above their heads so they had to knock on doors to ask what street they were in.

When he was still very young, Fred's family moved from the house by the railway sidings to one just next to Burnden Park, the home of Bolton Wanderers Football Club. Behind one goal the railway line ran along the top of an embankment and at the other end there were four tightly knit little streets packed with terraced houses built for the people who worked in the nearby Croft Lane Mill. The scene was immortalized in L. S. Lowry's painting *After the Match*, which was based on Burnden Park and its surrounding streets. The painting shows huddled figures in cloth caps, mufflers and clogs walking towards a football ground. The backdrop of smoking mill chimneys and terraced housing gives it the dark and forbidding atmosphere of northern industrial town life in the 1930s and 1940s.

Fred's younger brother Graham remembers the days just after they moved there. 'I think my earliest memory of Fred was when he was a kid, when we lived at home in Alfred Street. We had no bathroom suite or anything like that, not in the old terraced houses, so we used to have a tin bath in front of the fire. But Fred never liked getting in it and he was always screaming. So to keep him quiet, my mam used to give him blocks of soap, and he used to stick matches in this soap. Then he'd get bits of cotton to make ropes on the matches, as though they were going to be ready for putting the sails on.'

The main road from Bolton to Manchester ran past the end of the street. It was made of cobbled stones and it had two tram tracks running down the middle. Between half past six and seven o'clock every morning Fred was woken by the sound of clog irons on the paving stones as the workers tramped to the mills. All was hustle and bustle and the noise reached a great crescendo about ten minutes before work was due to start. Every Monday each street had its lines of washing hanging out and the coalman with his horse and cart was always in trouble with the local housewives if he had a delivery to make up their street on washing day. Trams rattled along the middle of Manchester Road and noisy diesel lorries lurched past carrying coal and bales of cotton to and from the mills. Steam engines were everywhere – in the mills, on the railways and, occasionally, trundling up and down the roads.

Fred never liked going to school, so his mother used to push him on to the tram that stopped near the end of Alfred Street. In the next mile and a half there were five chimney stacks along the way. Passing the mills, Fred used to peer through the windows of the engine house to get a glimpse of the gleaming leviathans inside with their great flywheels spinning round. And up above, there was the clag of black smoke curling out of the chimney when the fireman shovelled the coal into the boiler. Every so often long ladders painted bright red appeared on the chimneys. Fred always kept a lookout for them from the top of the tram and watched the men climbing up and down. Those fellows really fascinated him, but there were only two things he knew about them: they'd got red ladders and they wore flat caps.

Colin Shaw was in Fred's class from the time they started primary school and he remembers their early days. 'One of our first teachers was Miss Draper. She used to have what looked like a Hoover band round her head holding her hair in. As you went into the classroom there was a prayer area, and I can remember there was a little cradle. It had a red metal frame with two seats on and it rocked backwards and forwards. Right from his first day at school Fred seemed to develop an affection for this rocking cradle. But whenever he was on it he'd never let anybody else have a go.'

One morning, at assembly, the headmaster announced to the whole school that they were going to have some time off lessons to go out and watch a great big chimney fall. In strict single file they were marched from the hall to a field at the back of the school, and were told to sit perfectly still. The kids sat there and the tension began to mount as they waited for the giant to topple. They waited all morning but the chimney was still standing when they went home at midday for dinner. When they returned in the afternoon the chimney lay on the ground and they'd missed everything. Fred was about seven at the time, and he was really upset to have missed that chimney's fall. 'I never saw it come down,' he told his parents, 'I never did.'

But he didn't have to wait long to see another one as he grew up surrounded by relics of the Industrial Revolution. Much of the machinery and many of the buildings he saw as a child were just starting their decline and would soon be swept away. He understood at an early age that the world he lived in and the people he was seeing, like the steeplejacks and engine minders, had fascinating stories to tell and it was this that would shape his life and his character. Talking about it later, he couldn't recall exactly what had sparked this passion but he knew he'd had it from childhood. 'I can't remember when I first started noticing steeplejacks. My mother always used to say I was able to spot them from my pram. But whenever it was, this early fascination with steeple-jacking and steam never went away and I never really acquired any other interests.'

Fred liked the house in Alfred Street and he liked Bolton. In fact he liked it so much that he never wanted to go away. So he made a great hullabaloo every year when a trip to Blackpool was organized, for children with parents like his who couldn't afford to go to the seaside on a week's holiday. Fred would be there in his best gear and they'd try to force him on to the bus, but he wouldn't go. While all the rest of the kids went off to Blackpool for the day, Fred stayed at home watching the trains from his bedroom window.

'That was the thing with Fred,' recalls Colin Shaw. 'If he decided he was doing something, he was doing it. It didn't really matter who else it affected. Fred could be quite selfish like that. If things didn't go his way, Fred didn't want to know. If the teacher said we were going to do this, that or the other, and Fred didn't want to do it, he would sit there and put his head in his hands. The teacher would say, "Right, Fred, we're doing this," and Fred would just get up and do a runner. Me and another lad, because we were slightly bigger than the rest, used to have to go to the old shelter in the Rose Queen Field, at the back of the school, to fetch him back to class.'

The first teacher who had any real influence on Fred was Miss Buckley. Apart from doing the odd runner when he didn't want to do something, he was never a really naughty boy. But Colin Shaw remembers when he and Fred went off one lunchtime to have a look at some mill lodges. 'We got back to school late and Miss Buckley's punishment was to put us at the front of the class with our hands out. Then she got a ruler. We had the ruler on the back of our knuckles, which you'd probably end up in jail for now. But it seemed to have the desired effect on Fred because he was never really disobedient again after that.'

Fred was a well-behaved pupil but he didn't shine academically. There were about three of them who had problems with reading and they were eight years old before they could read properly. Fred always said he was no good at reading and writing, so they put him in the art group. They painted vases of flowers and bowls of apples, and when the teacher felt like it they went out for walks with their drawing books. Of course you couldn't go anywhere without seeing chimneys, so not only did he draw the chimneys he saw, but he put in red ladders and staging. 'What have you put those in for?' the teacher would ask. And Fred would answer, 'It was in my mind.'

When he was little Fred was quite shy. He was never one of the gang in the playground. 'We'd be playing football, throwing stones, chasing or playing tag, but Fred never ever joined in games like that,' says Colin Shaw. 'He was always a bit of a loner, standing in the corner. He'd either have two ball bearings in his pocket or he'd be playing with a piece of wire. He'd always be messing about with something. And, considering Fred lived next door to Bolton Wanderers football ground, I never ever saw Fred play a game in the whole of my life, be it football, cricket or whatever.'

His brother Graham agrees. 'When he was young he'd be different from the normal lads. When we were growing up the majority of the lads round here used to play football, but our kid and his mate, Alan Heap, would go down to the river, trying to kill rats with their catapults and things like that.' Fred himself said, 'I've always been a bit weird really; other kids would be playing Cowboys and Indians but I would be finding pieces of piping sticking out of the ground and be thinking what the hell is that? I must have had a mechanical interest even back then.

'Once I rode my bike down a narrow sewer. I was with my mate Alan Heap and when we got to the end we couldn't turn the bloody things round so we had to back-pedal to get out! That were an interesting tunnel where we could have been gassed but we never thought of things like that. The only thing I worried about were the fact that if it rained and you were playing around this tunnel – basically an open sewer – it would go from three feet of water at the bottom to over eighteen feet after a heavy downpour, and that were the scary bit.'

In the valley at the back of Burnden Park there were a few old mills and Fred and Graham would go fishing in the mill lodges there. 'Lodge' was a Lancashire word for mill pond and the lodges provided water for the boilers and also the cooling water for the condensing plants. Fred recalled that they supported an unbelievable amount of wildlife. There were goldfish and all sorts because the water was slightly warm. 'The other thing that was a bit sad really was that when I were a kid, there were quite frequent drownings of young children in the lodges. After a drowning they always filled about three of them in, so it wasn't long before they were nearly all gone.'

When I was with Fred at one of these lodges filming a piece for the *Industrial Age* series, he recalled a day he had spent with Alan Heap. 'Me and Heapy were quite experimental as kids and we decided we'd have a go at making a diving suit so that we'd be able to see what it were like on the bottom of the lodge. We got a Smith's crisp tin and a piece of motor car inner tube that we stretched round the bottom of the tin. And then on the top we soldered a piece of copper piping with a rubber hosepipe and a big lump of cork and a window in front. First of all we thought we'd try it out with a test dive in the local swimming baths, but when they saw this contraption they threw us out. So we had to try it out in the mill lodge and it were rather funny because Alan had to hold me head down, there were so much air inside this Smith's crisp tin. The thing is, it would never have worked anyway. You would never have got under water; you'd have floated along with your head on the surface with the amount of air that were inside the thing.'

Fred never learned to swim but, when he was a lad, water always had a certain fascination for him. The Bolton branch of the Manchester–Bolton–Bury Canal ran alongside the river Croal in the valley behind Burnden Park. The river and canal were both used as rubbish tips and Fred spent many profitable hours dredging with a rope and an iron hook for what he described as 'plunder'. This would all be hoarded in his mother's back yard or even in his bedroom because, he said, 'You never know when it will come in handy. But the river and the canal were not just a source of treasure trove, they were a place of adventure; a place for fantastic voyages.'

Alan Crompton was one of the gang and he remembers the things that Fred and Alan Heap used to get up to. 'Him and Heapy used to explore the canals a lot and one time they even built a canoe! They'd no money for materials, but they did have a few tools that Fred had inherited from his grandfather. They got about three or four sizes of bicycle-wheel rim; no spokes, no tyres, just the bare rim. Then they cut them in half and these became the ribs in the canoe. After that they got slate laths and they nailed these into the spoke holes. Then they got a wagon sheet and they draped it over this and nailed it on with clout nails. The whole thing was about fifteen foot long.

'They'd been building it for weeks, and it was in the back bedroom at Fred's mother's. The trouble was, when it was complete and the time came to get it out of the back bedroom, it was miles too big. So they ended up taking the window out and they slid it down the roof of the old shed at the back to get it down into the back street. Then they took it to the river Croal round the back of Burnden Park for a little cruise.

'The boat sailed all right, but it was very unstable. You had to sit dead centre in it or it would tip over. Fred mastered it and paddled a couple of miles along the Bolton–Bury Canal. But it took him a long time and his mother started to get worried. By the time he got home it was pitch dark, about half past nine. His mother went mad. She didn't know where he'd been and thought he could have drowned.'

'Our kid,' says Graham, 'and his mate, Heapy, were forever making rafts, and sailing them on the mill lodges, and we had really good times. We used to make dens in the disused mills and we used to go and sit in them and have a quick ciggie, where nobody could see you. We made one in the roof space of an abandoned factory once. We had a rope ladder that we had to climb to get in. And then, when we got up into the roof space, we'd pull this rope up so that nobody knew we were there.

'It was down by these old mills on the banks of the river Croal that our kid first got into steeplejacking. There were two big chimneys down there on this old bleach works. When the time came for demolishing the chimneys it was in the middle of the six-week summer holidays. Our kid would sit there on the bank of these lodges all day, from getting up in the morning till they finished work at five o'clock, and he used to watch the steeplejacks knocking these chimneys down. At the end of the day he'd say to my mother, "I'm going to be a steeplejack." And then Mother said, "You're joking, you must be mad."'

One of the stories that stuck in Graham's mind was of the time they were down by the river one Sunday afternoon, after it had been in flood. Fred saw some half-submerged pram wheels sticking out of the water and decided they were worth adding to his load of plunder. 'Our kid bent down,' recalls Graham, 'and started to trawl for it with a branch. As he did he whipped a piece of tin over to reveal fur boots and what was left of a leg. Further investigation revealed the leg was attached to a woman's body. Most of it had been eaten away by the rats because it had been there for quite a while. So we left it there and shot up to get help. The first place we got to was the works of a cough medicine manufacturer called Roberts Croupline, where we found the watchman and got him to phone the police. The police came and we were all there for quite a while. But they sent me away because I was the youngest.'

Fred's father, Frank, worked in one of the many bleach works that had grown up along the banks of the small brooks and rivers that flowed through Bolton. Like Fred, he was quite short in stature and, because of the damp environment he worked in, he suffered with his chest as he got older. Fred's boyhood friends remember he always had a cigarette in his mouth, but he never flicked the ash off. He left it burning and it was always falling off. He never really seemed to suck in any smoke. His only recreation was acting as a checker on the turnstiles at Bolton Wanderers Football Club on a Saturday, and cleaning out the hen pen on his allotment.

Bleach works had a lot of little steam engines that drove individual machines, whereas in the mills they had just one great big engine. At the place where Fred's father worked the machines were so noisy it was frightening, and they used to call the men who worked on them the ten loonies because, according to Fred, the noise of all the gears going round sent people deaf and daft. 'My dad used to come home from work,' Fred recalled, 'and, even though I would only be a baby, I remember he always used to be shouting and swearing about the job and those words ring with me for ever.'

When he got a bit older Fred had to go home from school at midday to take his father's dinner to the bleach works. Fred remembers it was an unbelievable place. 'Every corner you turned there was a steam engine throbbing away and driving some piece of shafting. They had big steam engines there for the main heavy bits, and little ones stuck in different places. More steam came out of the roof than went through the pipes, and every evening there were adverts in the local papers for stokers because the work was so bad they couldn't keep them.'

Fred's father worked at a contraption called a beetling machine. If you study a piece of cloth just after weaving, it's full of protruding fibres, and all of these have to be flattened. The modern method is to run the cloth through steam-heated rollers, but from the earliest days of cloth making, beetling machines were used, and there were still a lot of them in service in Bolton at this time. They were monstrous things, with cast-iron frames and rollers and beech beams. As the big cast-iron rollers with all the oak cogs turned, the cloth was dragged through the machine. And there were no safety guards around anything. Apart from all the dirt in the air, the noise was terrible and you couldn't hear yourself speak. No one can describe what it sounded like, although some literary character at the works called it Dante's *Inferno*. Fred's father spent forty uncomplaining years working one of these contraptions. Below the floor where he worked, there was a river that was alive with rats. They raced about below him and would climb boldly up to have a go at his dinner when Fred arrived with it. Two big chimneys above the bleach works belched out black smoke day and night. 'He was always surrounded by bleach and steam,' recalls Graham, 'and it was very cold and damp. No wonder he ended up with bronchitis, working in a place like that and smoking all the time as well.'

Fred's father hated steam. At the bleach works there were dozens of steam engines, driving hundreds of pulley wheels and belts and line shafting. There were no good jobs at the works, but there was one that was worse by far than all the rest. This was the job of the oilers and greasers: the men who went round with the oil cans and kept all the revolving machinery lubricated. The death rate among them was high, as they worked their way between the machines and the belts. Fred's father would come home many a time with a report of another poor oiler caught by his jacket in the machinery. He worked at that place from seven in the morning until five in the afternoon, and he'd come home dead beat. It was a terrible place, but the steam engines and all the ancient machinery fascinated Fred and he used to go down there as often as he could in the school holidays. His dad thought he was off his rocker to spend sunny days in this madhouse of industry.

Frank Dibnah's way of escaping from it all was his allotment and Graham has fond memories of the time he spent on it. 'Those were good days. We had a few hens and we'd the biggest rhubarb patch you've ever seen. It was ginormous, and the rhubarb was lovely. On summer nights, my dad used to come home from work and the first thing he did, after he'd had his tea, would be to shoot off up to the allotment till about nine o'clock. Then he'd come home, get washed and shaved and go off to the King Billy, over on the other side of Manchester Road, for a pint with his mates.'

Frank's other recreation was the football and the fact that their house was so close to Burnden Park suited him and Graham, who were both fanatical Wanderers fans. Fred never went near the ground, but he did make a few bob out of storing spectators' bikes in the back yard, and on the outhouse roof, at threepence a time. For a big game he could make thirty shillings in an afternoon, which was more than a day's wages.

Frank was working as a checker on the turnstiles on the Saturday afternoon in 1946 when they had a major disaster at the ground, and Fred's grandmother was actually on the embankment behind the goal where the accident took place. It was a cup-tie against Stoke City and there were too many people on the terraces. A lot of them had got in by gate-crashing where the ground backed on to the railway lines. They surged in from the back and the crush barriers gave way, crushing the spectators at the front, leaving a death toll of thirty-three fans.

When Bolton were not playing at home, Fred's dad

went to Bury to work on the gates there. Graham would often go with him. 'Then one day,' recalls Graham, 'our kid decided he wanted to come. It seemed a bit strange because he'd never been to a football match in his life; I don't think he'd ever even kicked a ball. But it wasn't the football Fred was interested in. All he wanted to do was go over the viaduct near our house on a steam train. When we got to Bury he just stayed on the railway station trainspotting until the football match had finished, and he waited for us to fetch him back home over the viaduct.'

School was still hard going for Fred; never very academic, he usually finished near the bottom of the class. His mate, Heapy, wasn't brilliant at school either, but they were both good at inventing things, and attempting minor civil engineering feats. They once dug a tunnel into a hillside, in the style of *The Great Escape*. They dug it into sand, shored it up with old picture frames off the tip and pieces of rotting floorboard and succeeded in driving it twenty yards into the hillside. Then they dug a shaft upwards and came out at the top of the hill.

Fred's next craze was to get a bike, because everyone else round his way seemed to have one. Even though his father slaved in a bleach factory all day long and his mother at the gasworks, the family never had much money, so a bike was something to be made from visits to tips. If you were lucky there would be a frame here and a couple of wheels somewhere else. After that you'd get hold of a pair of handlebars, and you didn't need to worry too much about brakes in those days. Building the bike was Fred's first bit of mechanicking. Once he'd got it done, it allowed him to get around a bit more. He found an old pair of binoculars and took them with him up to the top of the Earl of Bradford's slag heaps. From there, on clear summer nights, he could look out right to the edges of Bolton and see all the chimney stacks.

One summer night Fred spotted some strange-looking black lines around the top of a chimney. It was somebody's staging, so the next night he cycled all the way across town to have a closer look. As he got nearer, he saw red ladders with the name *John Faulkner, Manchester*, on their sides. Faulkner's was a well-established and well-respected steeplejacking firm who did a lot of the biggest jobs around Manchester and Bolton. From that evening onwards, Fred's hero was John Faulkner. He started to follow Faulkner and his men whenever he saw their ladders go up on a chimney. To Fred's mind they did things the right way, much better than anybody else. Their staging was always square, symmetrical and solid-looking, with ropes coiled up all nice and neat, whereas other firms seemed to have crooked-looking staging, with odd bits of rope blowing in the wind.

As Fred got older he began to stand out from the crowd. Colin Shaw remembers him as a likeable chap, who never did anything to anybody that could make you dislike him. 'If we were in a group and we all decided to do something that Fred didn't want to do, he would stand his ground and say, "No, I'm not going." He would start walking back home and find someone repairing locks on the canal or something and would start talking to them. He'd always stop and chat to someone, so if they were re-tarmacking the road, he'd start talking to the chaps with the tarmacking machine. He was always interested in things like that, where perhaps the rest of us weren't. We were just lads; we were there to play, but Fred to my mind was never a boy. In a sense he was always an adult. He didn't want to join in with the kids. In those early days Fred would just go and talk with workmen and give them a hand. He'd go to a forge or to a blacksmith's and just watch, and he would be learning all the time.'

Fred soon showed an aptitude for making things that was to serve him well throughout his life. He used to make model boats and he did a lot of wood carvings. Graham remembers him making a model of the *Titanic* one year for a national competition, which he won. He also made all sorts of strange vehicles and mechanical contraptions. 'He once made a soapbox with pram wheels and a tea chest and a handle on the side for a brake. Him and Heapy came racing down a street that ran on to Manchester Road. Just on the corner, outside the old pie shop, they were going so fast they couldn't stop and they went straight into the back of a number eight bus. But they got away with it unhurt and there was no damage to the bus. The soapbox didn't fare quite so well; the front wheels fell off and the whole thing disintegrated.'

Fred was getting more and more interested in the ladders he saw up the side of chimneys and he started to follow some of the steeplejacks around, studying the way they went about their trade. When he wasn't doing this he'd hang around mills to watch the steam engines, and look at how the machinery worked. The engines were connected to machines by line shafting, which is a complex system of belts and pulleys that transmits power from one engine to many machines. The steam engine that provided the power would usually be in a separate engine house next door to the main mill building.

Fred's knowledge of steam engines came from those days. He always said, 'It came out of being a bit of a burglar.' He lived near the bottom of a little lane that had a mill at the end of it and when business wasn't too good they shut down the mill and left it idle for four or five years in the hope that the good times would return and they would be able to start trading again. They left all the machinery and the engines in there as well. 'I remember,' said Fred, 'getting into the mill and lifting up a dust sheet to have a look at the engine they'd got under it. I found there was this beautiful British racing green steam engine under there with all the brasses left on. The boilers were still there and there was plenty of grease on everything, all ready and waiting for the next job. It was a bit sad really to see it standing there idle like that.'

All of these mills had big Lancashire boilers and the beauty of them was that you could burn anything on them. All the soot fell to the bottom and you could go for twelve months without cleaning them. Then, during wakes week, when all the mills in town shut down for the

week and everybody went away to Blackpool or
Morecambe for their holidays, they'd have to sweep out
the flues. For this the Black Gang would come in and
they'd have to get all the soot out of the tunnels. They got
a pittance for it and when the job was done they'd go to
the pub and get inebriated. During wakes week everybody had a reminder of how dirty the town really was. It
was the only week in the year when the skies were clear.
Blackpool was the mecca for the holidaymakers, but
some stayed at home in empty towns, while their friends
'blew in' a year's savings by the sea.

Fred knew that underneath every one of the chimneys
there was a boiler and an engine, so he plucked up
courage to wander into boiler houses and engine houses,
and he'd talk for hours, and listen. He had a gift for
listening as well as talking, and gaining friendships and
acquaintanceships, particularly with old men, old knowledgeable men, and they would give him things like
leaflets and brochures and old books on mill engineering
or steam engines or boiler-making.

Vantona Textiles was a massive firm in Bolton with
numerous mills around the town. Fred made friends with
the chief mechanical engineer for the firm and he used to
tell Fred tales about the engines that he had to repair,
including one he remembered mending in the middle of
the night. They'd been working on it from teatime and
they worked right through until half past three in the
morning, when they stopped and decided to give it a run
to see if it was all right. They opened up the valve and the

engine started going round and all the shaftings began working. The next thing they knew, all the mill operatives were outside waiting to start work. The thing was, in those days, no one had any clocks. They relied on steam hooters and the noises that came from the engines. So everybody in the immediate neighbourhood who worked in the weaving shed jumped out of bed to go to work as soon as the engines started because they all thought it must be half past seven, even though it was only half past three in the morning.

The man from Vantona also told him about a repair job on a great flywheel on a steam engine. Because of the amount of draught that was generated from the spokes as it went round, they boarded in the flywheels with rather beautiful tapered boards, wide at the rim end and narrow at the centre or axle end. To get at the big nuts and bolts that held all the sections of the flywheel together, they had to take off a few boards and get inside. When the job was done, they tightened up all the bolts, put the boards back and started the engine up. There was a great deal of rumbling and bumping coming from it and they thought somebody had left something inside. So they took the boards off to investigate and found it was the works cat. The poor cat was dead; it had been boarded in and had been going round and round with the flywheel.

Like the rest of the industrial north, Bolton also had a lot of engineering works and foundries as well as the mills. The foundries held a special fascination for Fred. 'They were dirty and unhealthy,' he used to recall, 'but when you think of the end product, like liquid metal – it's something else. I remember watching what went on in the blast furnaces at Hick Hargreaves. They'd spend three days loading them up with timber, coke, limestone and pig iron. Then the great moment would come when a man with a big iron rod unplugged the tapping hole and out would come a great stream of white molten cast iron. The molten iron would run into a big bucket for about ten minutes and then, when it was full, it would get hooked up on an overhead crane and swing right across the shop, before it was lowered on to an old cart that looked as though it had come off one of the first railways. Then they'd drag the cart with this great big bucket of molten metal in it out of the door and across the yard to the other side of the works. All very exciting stuff.'

2

Learning a Trade

Fred's schooldays rolled on, and he kept seeing more and more red ladders on the chimneys around him. His father's allotment was next to a greyhound track. It was on a site that had once been occupied by a colliery and the winding engine house chimney was still standing. Like a lot of these old chimneys, it wasn't serving any useful purpose and because of the state it was in it was becoming a liability. So the chap who owned the site decided it was time to get rid of it. The men who had been hired to do the job arrived with ladders that had once been red but were now a wishy-washy pink. They began the job by laddering the chimney and knocking the top off it. After that they took the ladders down, chopped a big hole in the bottom of the chimney and put blocks of wood into it. Fred didn't know what was going on, but it didn't matter because he had a ringside seat in the allotment with his father.

It was a gloomy Sunday in the middle of winter when the demolition men lit their fire under the chimney and the flames started to dance up in the dark. The site owner put all the lights on around his greyhound track, which made it all quite dramatic. Excitement began to mount – and then things went badly wrong. Not many chimneys were knocked down in those days and the demolition men didn't really know what they were doing. So when the chimney came down, it didn't go anywhere near where it was supposed to. On the site there were some dog kennels and next to them a small café. The chimney stack came down right on top of them, cutting the dog kennels in half and chopping a corner off the café. As it also ripped down all the power cables, every light went out. It was a sight that Fred never forgot through all his chimney-felling days.

At school Fred's dismal academic record didn't do anything to bolster his mother's hopes for him. For Betsy, Fred was going in the wrong direction. She wanted him to have a clean-hands job and at one time it looked as though he might, when his interest in steam got him into art school. The entrance examination was in three parts. First he had to draw a bunch of flowers, then came a perspective exercise, and then he had a free choice of subject. Fred drew a steam engine, based on a magnificent traction engine that belonged to a showman he knew. He used to go to the fair every year just to look at it and, to his mother's surprise and delight, it was the drawing he did of this engine that got him into art school. But when he was there, his persistence in painting only industrial scenes soon put an end to her idea that he might actually become an artist.

In later years Fred would always speak of the benefits he got from his art education. 'I'm glad I went to art school; drawing makes you more observant. You see things and you notice detail, and you see beauty in all sorts of things.' Being able to draw also turned out to be useful for his business. His ability to illustrate things with clear drawings tended to win him jobs in competition with less artistic rivals, who would turn up with lopsided scribbles on the back of old pieces of wallpaper.

Fred soon realized, though, that to get anywhere as an artist you've got to be either brilliant or very lucky, and he was neither. His drawings had a style of their own. He was never very good at drawing people so he stuck to industrial themes: machinery, pithead gear and spinning mills at night with all the lights burning, factory chimneys with scaffolding round the top, canal scenes, landscapes of a world that was fast disappearing. But he enjoyed his three years as an art student. 'The short period I went to art school, I'm glad I did that. It didn't do me any harm because we used to have forced summer outings with the drawing board and paper and pencil to look at things that were of note in the building line. To learn about perspective, and be able to draw a building so that it doesn't look like it is going to fall down, is very difficult for a lot of people, but I find it quite easy. I have no problem whatsoever in drawing great big tall factory chimneys that look like ... great big tall factory chimneys!'

After art school Fred thought for a time that he might end up with a clean-hands sort of job, just as his mother had hoped. But he'd set his heart on being a steeplejack. The trouble was there were no vacancies for would-be steeplejacks when he left school, so he went to the youth employment bureau, as it was called in those days. The chap behind the counter said, 'Well now, we have one very good job available. It's an assistant at an undertaker's.' So Fred was sent to a funeral parlour, where he thought the undertaker looked like a corpse himself. The man opened a door and showed Fred into a room filled with trestles and coffins. Fred didn't like what he saw and straight away he thought, this isn't for me. So he acted as though he wasn't very bright and the undertaker sent him back to the youth employment bureau. 'Right then,' said the man behind the counter, 'we've got another good job available at a joiner's. The trouble is six other lads are after the same job and they're already on their way there.'

'I've got my bike,' Fred said, and he got there in double-quick time. The joiner was called Mr Rawlinson. He was a nice man, and by way of a test he asked Fred if he knew what a tenon saw was. When Fred told him his grandfather had left him one, it went down rather well and Mr Rawlinson immediately offered him the job, before the other six hopefuls had arrived on the scene. Fred was sixteen.

Bill Rawlinson's daughter, Valerie Taylor, remembers her first impressions of the young Fred Dibnah. 'My dad's firm usually had an apprentice working and another man as well, sort of an older joiner. It was only a small firm, really only three men including my dad, but he had his own car so he wasn't doing too badly because in those days not many people had cars. When Fred started I thought he was quite a good-looking youth. He had a cheeky grin and was a bit of a Jack the Lad and he was a real practical joker. But he was a very good-hearted lad; he would do anything for anybody. Even then he wore his flat cap because he always saw the steeplejacks wearing flat caps as they were climbing up the chimneys, so that's what he wanted to wear. He was also very useful to my father. My father thought a lot of him and my mum did as well – we all did. Anything he was asked to do he would always tackle the job very willingly. Fred got on very well with Dad and when he'd worked there for a bit Dad asked him to call him Bill. So Fred did. He used to call him Gaffer or Boss before that. I used to go to the workshop with friends quite a lot in the summer holidays and we'd just sort of mess around in the shavings that had come off the saws. The smell was nice; I've always loved the smell of wood shavings since.'

Valerie's sister Barbara remembers what Bolton was like when Fred first started work for their father. 'Bolton was very different then. You'd still got loads of industry and it was certainly a lot safer than it is now. You could go out at night and not have to worry about what was happening. But it wasn't as healthy as it is now; we had loads and loads of smog through the winter. We lived on

Chorley Old Road and the buses used to come up and we couldn't see the numbers on them because the fog was so thick.'

'All hustle and bustle' is the way Fred's friend Alf Molyneux remembers it. 'You went out to work in the morning, walked to the bus stop and two buses would go past full. You'd no chance of getting on them. Of course there weren't as many cars about then. But it was very busy; a time when you could knock on any door and get a job and they'd say, "Can you start now?"' Another of Fred's friends, Jimmy Crooks, was starting work as a coal miner in the nearby town of Wigan round about the same time, in the 1950s. 'Everybody had clogs on as they went to catch the six o'clock bus from the centre of Wigan. Fifty or sixty men all walking down with clogs on must have woken everybody up if they weren't already up and on their way to work themselves. The mill girls used to start at six as well. It was like the middle of the day, especially on Market Square where all the buses were. There would be hundreds of people knocking about. Not like now, they're all still in bed.'

In the early 1950s the Lancashire coal industry was still in full production. Coal was needed for the mills because most of them were still operating on steam power. But the decline of the cotton industry had already started. Wartime disruption of world trade routes had encouraged the formation of textile industries in India, Japan and China. And exports of British machinery ensured that these indigenous industries grew rapidly. Lancashire was finding it increasingly difficult to compete on price in these, its traditional export markets. During the 1950s electric power began to replace the old steam engines and the dingy working conditions in the mills started to improve. But the industry could not afford to invest in the latest equipment. Instead the mills responded to demand by using ancient machines and running them flat out. By the end of the decade, foreign competition began to hit home markets as cloth was imported from Hong Kong, India and Pakistan. But for Fred there was plenty of work with Mr Rawlinson all through this period.

One day, soon after he'd started the job, Mr Rawlinson asked Fred if he'd ever done any pointing. He'd got a new window frame to put in, but the owner of the house needed some pointing done and he didn't have a clue about working with mortar. Fred said he'd have a go and, after doing the window, Fred had to point the whole garden wall, which took him about a fortnight. The old woman who owned the house was so impressed, she asked if Fred could do the same to her gable ends. They were three storeys high, and Fred didn't have the tackle he needed to get up there. Mr Rawlinson took him to one side. 'I don't want to turn down work,' he said. 'Can you do it?'

'Oh, it's a very basic steeplejacking job,' Fred assured him, 'and I've studied all those techniques. I can lash all your short ladders together with a bit of a rope and a bit of hardboard in between.'

'How Fred managed it with his short ladders and bits of

board lashed together, we never knew. But he did manage it,' Valerie Taylor recalls. 'The lady was delighted with the job, and so was my father. Fred was pleased with it as well because it had given him his first bit of steeplejacking experience, so everybody was happy. And my father realized it was an extension to his joinery business. He now went in for what he called property repairing, with Fred pointing one gable end after another.'

Fred enjoyed working for Mr Rawlinson. He was a good man, but like most others in those days he didn't have very much money. What he did have was an old car. It was a Wolseley Hornet with a leather top and what they used to call a sunshine roof. These things were good when the sun was shining but they did have a tendency to leak in the rain. Eventually Mr Rawlinson filled in the holes with wood shavings, put a sheet of felt over the top, and tarred the whole lot. That was high technology for those days. He also had a trailer with old Austin wheels from the 1920s. Mr Rawlinson never once painted them, and the rust had been eating into them for years. Although each wheel was made of two metal plates welded together, you could see daylight through both of them. When they'd got a job to do, Fred loaded up the trailer and Mr Rawlinson drove him to the site. Then Mr Rawlinson left and Fred had to find his own way back to the yard.

Fred used to get quite worried about the trailer, particularly when it was overloaded. Sometimes he would fill it with six cubic yards of solid wood, all wet through from the yard. One dark winter morning he loaded up the trailer and Mr Rawlinson set off for Ashton-under-Lyne, which was about fifteen miles away. Normally he would have been back about half past five, but when Fred locked up at six Mr Rawlinson had not returned. When Fred got to work the next morning the Wolseley was there but no trailer. Mr Rawlinson told Fred that he'd been driving up the road from Bolton to Bury when one of the wheels finally gave up the ghost. The rim with the tyre on had sailed past him while the trailer, with its load, ran on to the grass verge.

'What now?' Fred asked him.

'Well, I managed to find another axle with wheels on it,' he replied. 'It's going to be a bit of a complicated job though, but I'm sure you'll manage to get it done.'

And sure enough Fred did.

All the time Fred was working for Bill Rawlinson he would study the nearest chimney, particularly if it had red ladders on it. He still wanted to be a steeplejack in a flat cap, going up and down those red ladders. But Fred was employed by Mr Rawlinson. By this time Fred had been with him longer than anybody else, though every now and then if they were a bit busy someone else would be signed on to help. One day a strange sort of fellow came along. He smoked a pipe and he always wore a long black over-coat. Mr Rawlinson called Fred over: 'I want you,' he said, 'to keep an eye on that fellow. He likes a drink, you know.'

Nothing unusual about that, Fred thought, but he said nothing.

There was an old handcart in the yard for doing local jobs. Its condition was not much better than the trailer. One morning the new man and Fred loaded it with two lengths of guttering, two ladders and their tool bags. They were about halfway up a hill when the new fellow took a hammer out of his bag and began to smash one of the wheels. The cart was pretty rickety and it fell over. Fred stood there inspecting the damage. There were no mobile phones in those days so Fred had to go to the nearest telephone box and ring up the gaffer to say the old hand-cart had fallen to bits. This meant the gaffer coming out in his car and trailer. Mr Rawlinson didn't seem too surprised about the handcart, so Fred kept his mouth shut about what had really happened.

It soon became clear that this new fellow really did have a drink problem. But Fred continued to keep quiet. Each afternoon the man would say to Fred, 'Just going for a box of matches.' Then he'd be away for an hour or two. Once Fred was alone on a roof when he heard the ladder shaking and clattering. It was the new fellow, and he was properly tanked up. He was in such a state that Fred had to send for Mr Rawlinson to help get him down to the ground, and that was the end of him as a workmate. Fred was becoming a trusted worker – an indispensable part of Bill Rawlinson's little organization. As Valerie Taylor recalls, 'My dad always had a lot of time for him. He was always put in charge even though he was much younger than some of the other men.'

Soon another new fellow came to replace the drinker

and right from the start Fred got on well with him. One day they were on their way to a job when they saw an enormous chimney stack out in the country. It was 262 feet tall, had the familiar red ladders up the side, and immaculate-looking staging round the top.

'I'd love to climb up there,' Fred said.

'You'd never manage that,' said his new workmate.

So Fred bet him ten bob (50p) that he would not only get to the top but also leave proof of it. The job they were on their way to was in a park. When they got there Fred found a large Union Jack stored away in the park keeper's hut.

What perfect proof, he thought, and without saying a word to anybody he rolled up the flag and shoved it inside his jacket. On his way home from work that day a young roof slater he knew spotted the bulge that the flag was making under his jacket and asked him about it. When Fred told him what he'd nicked it for, the slater said, 'I've got a Union Jack at home. I'll give you mine as well.'

Fred had noticed that there were two lightning conductors on top of the chimney, and now he'd got two Union Jacks – one for each of them. It was time to earn his ten bob.

The climb would have to be done at night because Fred didn't want to be seen, and the real steeplejacks would be working on the chimney during the day. It was situated a bit of a distance from the mill itself and had a long flue that ran underground through a small wood. When Fred got close to the stack he found he had a little problem. The steeplejacks had taken away the bottom ladder to keep the foolhardy from attempting a climb, so there were sixteen feet of chimney to get up with no ladder. But Fred wasn't going to be put off by a small thing like that. He tied a big nut to a piece of string and threw it over the bottom rung of the second ladder. Then he used the string to pull a rope through and with this rope he hauled himself up commando-style and began his ascent. When he reached the top he tied the two Union Jacks to the lightning conductors, came down and went home.

When he got home there were lots of questions about where he'd been and what he'd been up to; but he managed to evade them and put his parents off. Next night, in the evening paper, there was a big picture of the chimney with the two Union Jacks flying proudly on the top. But Fred wasn't going to get any glory for it because the authorities claimed it had been done by Manchester University students, who were celebrating their rag week. Valerie Taylor remembers her father coming home from work that night and as they were all sitting round the table having their tea he read the paper and said, 'That's young Freddie. That's not Manchester University. There's only one person who could climb that chimney and that's young Freddie.' But because of the newspaper report Fred had quite a job collecting his ten bob, until the slater backed him up about the second flag. 'He would have been about seventeen or eighteen at the time and he just had no fear,' Valerie recalls. When Fred went into work the next day Bill Rawlinson said to him, 'Get it out of your system, Fred.'

Fred's opening move in the chimney side of the business came when he decided to build one on his mother's house, which he had to sweep for the rest of his life because nobody else would. We went to the house with Fred to film an introduction to a programme on building houses for *Magnificent Monuments* and Fred told us about one of the first bits of building work he did. 'I built this chimney when I were about seventeen years old,' he recalled. 'We had a stack like the one next door, with five pots on, but four of them were disused and smoke only went up one. That was the one for the back kitchen fire and the back boiler. The other four just dripped water down the walls of our bedrooms and next door's.' So good old Fred decided that he would take the chimney stack down and build a new one for his mother. 'I set off with brand new Accrington bricks and there were various comments in the neighbourhood like "What's he doing?" because it was a bit unusual to see a seventeen-year-old lad knock his mother's chimney down and build another one. I didn't have any real design or plans for it, but as I built it I kept getting another idea about what I wanted it to look like, and when it was completed it turned out to be a most handsome thing. Everybody said I were crazy, but it's never cracked and it's got a lovely draught on it; suck your house slippers off when it's going at the bottom! It has stood there now for some forty years, and when my mother died my brother and I sold the house. The people who live in it now decided they wanted to knock it down, but the local authority put a preservation order on it. So it is still there till this day, and looking just as good as the day that I built it.'

But Graham Dibnah remembers it all a bit differently. 'The chimney at my mother's house is a bit of a sore point, with me being a bricklayer. That old chimney we had was dilapidated, falling down, and I just happened to say to Fred one day, "Hey, kid, that bloody chimney looks in a bad way. Why don't we take it down? We don't use four of the pots, so we can slate them over and just have the one from the fire downstairs."

' "Ah, a good idea, that, kid," he said.

'So anyway we got some bricks and knocked it down, and then started building a new one. We got it to about two foot six high, out of the roof, and Fred said, "I think we'll alter it. We'll taper it in." So I let him have his way but I told him that I wasn't going to have anything more to do with it. So he built it higher and tapered it in, so that at the top it was only four inches across, which was no good at all. Anyway, after about three months the fire wouldn't draw properly. It was all choked up with soot. So what he did then was to climb on the roof with a sack and some half-bricks, and a rope on the end of this sack, and he used to put this rope down, getting rid of all the soot down in the bottom. But one day he was doing this and the bag fell off the end of the rope. It came crashing down, went through the lead pipes and burst them and we finished up with about three inches of water all over the kitchen floor in no time. I can remember my dad going mad with him about it.'

The next idea that Fred came up with was to get hold of a dummy and take it up to the top of another chimney so that it would look as though somebody had hung himself there. Mr Rawlinson always went home for his dinner, so Fred started to make the dummy at work using wood laths and fencing wire. It was basically one upright stick for the backbone and another stick across for the arms. The head was made from a sack full of old rags. The next job was to dress this dummy. Fred got hold of some old trousers and a pair of his father's boots, which he filled with concrete to make the thing hang right. Then he got an old overcoat, which his mate's wife had got from a jumble sale. The whole thing folded over so that Fred could tie it up in a sort of haversack to get it up the chimney.

The stack that Fred selected was 212 feet tall, and it was right near the town centre. John Faulkner's men were working on it with their red ladders all in position. Again Fred went there at night, but all the factory lights were on and the boilers were going full blast. The climb up to the top was no trouble for Fred and when he reached the staging it was pitch black. But it was a windy night and, as soon as he started to untie the dummy, the wind got inside it and the whole thing started to flap around like a mad bird. As he fought against the wind, over two hundred feet up in the sky, Fred realized that his mission wasn't going to be a great success. The dummy was being blown all over the place and Fred couldn't get it to hang right. In the end he had to give up his battle against the wind and leave the dummy sticking out from the chimney at a forty-five-degree angle. Fred came down the chimney, bought himself some fish and chips and went home.

When the next day dawned it was foggy and there was no wind. It was Saturday and, as luck would have it, Fred was working right over on the other side of town. As soon as it was knocking-off time, he left the job he was on and headed for the chimney. There was no sign of the hanging dummy, but groups of people stood around on the street corner as though there had been some sort of disaster. Fred went up to a couple of women and asked them what was going on. 'Some silly sod hung a dummy from the top of that chimney,' one of the women replied. Then the women told him that an old boy who lived opposite the mill had woken up that morning, looked out of the window and seen the gruesome sight. He rang the police and before long the whole neighbourhood was up as police cars, fire engines and an ambulance all came rushing to the scene. The mill manager had been dragged out of his bed. Then they got John Faulkner's foreman down to the scene. The police told him he would have to climb the chimney and bring the dead body down. But then someone arrived with a pair of binoculars and they realized it was a practical joke. When Faulkner's foreman brought the dummy down they found a Manchester bus ticket in the pocket of the jumble sale overcoat, so the university students got the blame again.

A few weeks later Mr Rawlinson had Fred fit a new floor in a semi not far from the chimney. The lady of the house was very talkative, so Fred asked her if that was the stack the Manchester students had hung the dummy from. To Fred's amazement, she broke into a frenzy. 'It may have been a dummy,' she said, 'but my husband didn't know and he had a heart attack. Fortunately he got over it,' she continued, to Fred's relief, 'but it were a bad sort of thing.'

Fred's chimney-climbing antics had got him thinking. There he was, just turned twenty, and after four years he was still doing the same old joinery and pointing jobs for Mr Rawlinson. True, the boss let Fred call him Bill, and Fred was his senior man; but Fred was neither a steeplejack nor his own gaffer, both of which he wanted to be. Whenever he spoke about this to his dad, Frank would say, 'You're still young yet, Fred, something will turn up.' And soon something did turn up. Fred was walking along the bank of the river Croal one day when he spotted a ladder floating in the water. The river was more like an open sewer in those days but a ladder was a ladder, so Fred fished it out, cleaned it up and painted it bright red. This ladder enabled him to start doing some jobs for himself. After pointing people's houses all day for Bill Rawlinson, he'd point other people's houses in the evening and at the weekend.

Any time Fred was sent out by Mr Rawlinson with the handcart to have a look at a job, it would take him an hour or two to get to where he was supposed to be going, even though it might be just around the corner. The delay came about because, whenever he saw a chimney that was laddered, he would always go and talk to the steeplejacks, picking their brains and finding out how they put up their ladders and staging. Many a time the clients expecting him would phone up his boss. 'I thought you'd sent somebody to do our job. Where is he?'

'Oh, he'll be there, don't worry, he'll be there.' And then Fred would arrive about an hour and a half late.

At weekends Fred was getting more and more work on houses. He pointed them with mastic and linseed oil, and saved all the money he was earning so that he could buy some ladders. Eventually he earned £90 and he was able to buy five good ladders. After that it was a case of getting hold of pulleys and ropes; and Fred knew where to go for them. There were a lot of pulley wheels left lying about in half-dismantled mills. As for ropes, the best source was near Burnden Park, where there was a railway goods yard and a great iron bridge that crossed the river Croal and the Bolton–Bury Canal. Kids used to take ropes from the yard to make swings under the bridge, so at night Fred would go down there and help himself to their swing ropes. And all the time he continued to go round watching the steeplejacks at work to see how they did it and to talk to them and learn what he could from them.

One of the steeplejacks that Fred used to watch in the 1950s was Tom McFarlane. Tom had started off steeplejacking in the 1930s when he was sixteen. 'Fred Mills wanted a lad to start with him and I went round and got the job. It started Monday morning and I went right up to the top of a chimney he was working on. It never bothered me; I took to it like a duck to water. I had never been up high anywhere before. At that time there was plenty of work around here, but it was not till after the war that the boom really started. All the mills had been working flat out all through the war. They'd not had time to get maintenance work done, but they'd also accumulated a lot of money, so they all had something to spend. All the chimneys were in really bad need of repair. Because of the state of them there were a lot of fatalities round Manchester and quite a few lads died.'

Tom McFarlane worked for Faulkner's for four or five years. As an old-established firm with a very good reputation, they got all the best jobs around Manchester. One of the secrets of their success was that they made all their own mastic and they could guarantee that it would last for forty years. Tom McFarlane remembers Fred watching him at work when he was a teenager. 'It was his story about the Barrow Bridge chimney that he'd climbed to put the flag on that made us become aware of this lad. Fred was very friendly and you could have a good chat with him. He was really interested in anything we were doing, and vice versa. I didn't see him as the competition; I was getting a bit too old for steeplejacking then. But I think Fred went on to be the greatest steeplejack ever.'

Around this time Bill Rawlinson landed a job for Fred at a big mill. The job involved putting in a baling press, and he had the run of the place while he did it. The mill had six floors, and a tower that was badly in need of repointing. One day, when Fred was talking to the chief engineer, he mentioned this casually to him.

'Aye,' the man agreed, 'I've heard you do a bit of that sort of work. How much would it be for the tower?'

Fred was desperate to get the job so he came in with a ridiculously low price of £140. Then he had to explain that he could only do the work at weekends. Luckily the engineer was prepared to accept that, and Fred used every spare minute of his time repointing the mill tower. By then, Bill Rawlinson had realized there was more to Fred's interest in steeplejacking than he had at first thought. In those days the fatality rate among steeplejacks was pretty high, and when one fell off a chimney it was always mentioned in the papers. Whenever that happened Bill would cut out the article and pin it up on the workshop door. It wasn't long before the door was covered with pictures and details of dead steeplejacks. 'It could happen to you,' he used to say to Fred.

Valerie Taylor remembers those times well. 'When Freddie kept talking about being a steeplejack, my dad was a little bit upset. He didn't want to lose him, didn't want to lose a good workman. So he did try and put him off by pinning those little notices on the back of the door. But it didn't deter Fred at all. He wanted to be a steeplejack and that was his aim in life.'

But doing the job on the mill tower didn't bring Fred any new openings. Although he had watched Faulkner's men many times, he still didn't really know how to put ladders up a chimney. He did talk to Faulkner's men but a lot of them weren't very bright. Most had started as labourers, just doing what they were told. They had learned the job as they went along – first to tie clove hitches, then to pull ladders and planks up chimneys – but it was all a bit vague and they were not very good at describing what they did. The big breakthrough came when Fred met a steeplejack who had started his working life as a draughtsman. This chap had only come into steeplejacking to help his father-in-law out and he was able to explain everything to Fred in simple drawings. He showed Fred how he could get himself up a two-hundred-foot chimney without killing himself.

But there was another problem. Although Fred had got some fancy cards printed with *F. Dibnah, Steeplejack* on them, whenever he went into the front office of a big mill or factory, people looked on him as some sort of little lad and he hardly got anywhere. The only jobs he could find were painting little tin chimneys on the back of dry cleaners' shops. They were insults really, being only about thirty feet tall. But Fred had his eye on something special that he felt would get him established in the steeplejacking business. There was a chimney to be demolished on a particularly nasty site, which other steeplejacks avoided like the plague. It was right on the side of the main railway line from Bolton to Manchester. There was a signal box just near it and in those days signals were operated by dozens of steel wires running over hundreds of cast-iron wheels. It was a very difficult steeplejacking job and, if you were to make one mistake, it would have put the whole railway system out of action. Fred was determined to get the job, but selling himself was a real problem.

Then he had another breakthrough. One day, as Fred was walking through town, he bumped into one of his old teachers from the art college, Lonsdale Bonner. Mr Bonner ran a very impressive-looking Jag, but, even more importantly, he could talk posh. In spite of this he never had much money. When they met, Fred told him all about his steeplejacking ambitions and how they were being thwarted because he couldn't get past the front offices of the mills and factories. So Fred and his old teacher struck a bargain. For every job Mr Bonner got him, Fred would pay him £10. Fred told him about the horrible chimney job next to the railway line and how much he wanted it. Mr Bonner went to find out about it and with his big Jag and posh talk he got the job for Fred.

Now Fred had to prove himself. For years he'd watched the other steeplejacks and talked to them. But this was the first chimney he would be doing himself. He got his ladders and the rest of his tackle out of the yard and set off for this fearsome stack. He laddered the chimney and put the staging round the top with no problem. All those years of watching and learning were paying off. He then pro-ceeded to take the chimney down brick by brick, starting from the top. So he earned £100 the hard way, and Mr Bonner got his tenner as agreed. Graham Dibnah remembers that first chimney well because it wasn't far away from his mother's house, where they were all still living at the time. 'Out of our kitchen window you could see right up the back street, and you could see this chimney, and my mother was stood at the sink one day and she said to my dad, "Look at those silly buggers up there." She didn't realize at the time that it was our kid who had got his first job.'

Eventually, after a few more successful jobs, Fred started to find some for himself. A good place to get work was the local temperance bar, which was run by his uncle Fred. It was over the other side of Manchester Road, opposite Burnden Park, on the corner of the back street where Fred had demolished that first chimney. Uncle Fred made his hop bitters in a wooden barrel, stirring the mixture up with a big wooden spoon. Then there would be a special bottling day, when Uncle Edward, who was a signalman on the railway, helped to decant the brew into bottles. They'd crate the bottles and then stack them up to allow them to ferment. On Sunday mornings Uncle Fred's temperance bar was packed; people went there and drank a pint or two of this non-alcoholic brew to warm them up for the real stuff when the pubs opened at twelve o'clock.

Fred and Graham and all the lads from around Burnden used to congregate there in the evening to drink Vimto and sarsaparilla and a non-alcoholic stout Uncle Fred used to make. Upstairs was a billiard hall and every night it was packed with lads playing snooker. They didn't have any jukeboxes or gaming machines in the temperance bar, everybody just chatted. After they'd finished training, a lot of the Bolton Wanderers footballers used to come in to have a game of billiards or snooker. Tommy Banks, an England full-back, used to be a regular. He always caught the bus outside the temperance bar, back home to

Farnworth where he lived. How the world has changed for a professional footballer. Can you imagine a footballer today going into a temperance bar, drinking Vimto then catching the bus home?

Fred wasn't bothered about the footballers but when he was looking for work he would go to this bar all blackened up, to let everybody know he'd been doing something worthwhile. One day a fellow came in who appeared to be a cut above the rest, mainly because he was wearing a suit.

This chap showed Fred a chimney that you could just about see from the window of the temperance bar. It was clearly in a pretty poor state and Fred said the whole thing could do with mending and repointing.

'Why don't you go and see them?' the chap asked.

'I can't go in there,' Fred said. 'It's all a bit too posh for me.'

But the chap in the suit said, 'Well, as it happens I know Mr Farnwood, who owns the mill. I'll have a word with him for you.'

The next Sunday morning Fred was at Uncle Fred's temperance bar again and the chap with the suit came in to see him.

'I've seen Mr Farnwood,' he said, 'and he'd like you to go to the mill with a drawing of what should be done.'

Because Fred had been to art school, doing the drawing was no problem. He did an illustration of how a lightning conductor should be fixed to the chimney. The drawing had plenty of detail in it, and it was all nicely coloured.

The main thing, though, was that it impressed Mr Farnwood. But there was one condition.

'What's that?' said Fred.

'You can have the job,' Mr Farnwood replied, 'and if you do it well I'll pay you. If you make a mess of it, you don't get paid.'

'Fair enough,' Fred said. All that mattered was that he had got an important job at last. And there was no way he was going to make a mess of it.

Fred laddered the chimney, and then he mended the lightning conductor and repointed some five feet of the brickwork round the top of the stack. Mr Farnwood was happy with the job so Fred got paid. And he got another job out of it, repairing the roof of the factory. While he was up there Fred had a very useful find. The ladders he'd acquired so far were fine, but his planks had come from the building trade. This meant that they were too short and narrow, not really right for steeplejacking. But up on the factory roof Fred found some longer and wider planks that were used to get across the roof lights in the weaving shed. There were so many of these good planks that Fred did a swap, leaving his short narrow ones up there in place of some of the wider ones.

Weeks later he got a letter from Mr Farnwood himself, asking Fred to come to his mill immediately. This is it, Fred thought, he's missed the planks that I nicked. When he got there, though, Mr Farnwood took him to one side and said, 'See that other chimney over there with that steeplejack working on it?'

'Yes, sir,' Fred said guardedly.

'Well, I don't like what he's telling me about it. I want you to go up and have a look at the job yourself. Then I'd like you to give me your opinion on it as if you were my consultant.'

Fred wasn't very happy about this. In truth he felt it was a bit unethical to spy on the work of another steeplejack. But because of those planks that he'd swapped, he thought he'd better do as he was asked. It was a big job, and in the end Mr Farnwood had Fred doing it, with this other fellow as his assistant. So instead of being handed over to the police for nicking the planks, Fred made good money on that chimney, and with it he bought himself some more ladders. But he was still living at home and as his parents' back yard was only fourteen feet square it was getting a bit overcrowded. Fred had got his red ladders stacked halfway up the side of the house and his father began to get annoyed about this because his mum couldn't see out of the windows.

Around this time Fred did a lot of work for the Earl of Bradford's estate. The agent for the estate, Stanley Dalglish Boardman, who gave Fred the work, was, as Fred remembered him, a bit of a tyrant. Everybody feared him and on the earl's slag heap there were notices in white enamel and dark blue lettering: *People found picking coal on this waste will be prosecuted. Stanley D. Boardman, Earl of Bradford's Estate, Bolton.* The earl's estate had a storage yard and across the street was another yard belonging to a painter and decorator whom Fred knew, so Fred did a deal with him.

The decorator was looking for someone to do some repairs to his premises, while Fred wanted somewhere to put his ladders and other tackle. The premises were in a terrible state. The wooden fence at the front was falling down, and he had to lift the gates to move them. As for the sheds themselves, they were made out of old railway sleepers and bits of timber. Fred built a new brick wall along the front, rehung the gates and sorted everything out inside for his ladders, planks, ropes, pulleys, hooks and brackets. By this time work was good. Fred was getting busy and at last he was beginning to make some money.

But it wasn't all work. Graham Dibnah remembers what it was like living in Bolton in the late 1950s. 'Oh, Bolton was a brilliant town when we were lads. We used to enjoy it. There were still a lot of mills. And some of the girls our age who worked in the mills lived down Croft Lane. They used to do shift work; there was a six/two and a two/ten shift and sometimes they would be on days. At the end of the shift they'd come home past our house with white socks on, all covered in cotton, and their clogs and their pinnies. They were always friendly and we used to talk to them all at the end of our street. I used to go to night school then for my trade as a bricklayer. We finished at nine o'clock. We'd go and meet the lads at the pub and then we were off to Bolton Palais. One and sixpence, I think it was, jiving and rock and roll, Tuesday and Thursday. Then we'd go on Saturday night and that was the night for ballroom dancing. But we had some good times, believe me. Fred used to go as well, but he didn't go jiving. He couldn't rock and roll to save his life. He only went on a Saturday, to ballroom dancing. But even then he didn't do much dancing. He was more interested in having a look at the women.'

One of the girls who used to go down there was Bill Rawlinson's daughter Valerie. 'Bolton was a wonderful town to grow up in. We had so much. We had about fifteen cinemas in the town, lots of dance halls, the Albert Hall which had all the big bands coming in on a Saturday night and on a Sunday night teenagers could go and have a sing-song, and we'd lots of coffee bars.'

Although Fred's steeplejacking business was beginning to take off, he still had his day job with Bill Rawlinson. During the weekends and evenings he worked all the more for himself. The name Fred Dibnah began to be known wherever there were tough jobs to be done, especially if they were high up. Fred became able to plan ahead. Then the army and National Service caught up with him.

By this time Fred was twenty-two years old, and he should already have started his National Service. But Bill Rawlinson didn't want to lose his best worker. 'My dad was quite eager to keep him out of the army,' said Barbara Yates. 'Years ago, if you had an apprenticeship, you could stay out of the army until you'd finished it. My father managed to keep Fred out until he was twenty-two, but then he had to go to a tribunal in Manchester where he tried to argue on Fred's behalf, saying that he was far too valuable to go into the army. The selection board thought about this, conferred, then said to my father, "We are very sorry, Mr Rawlinson, but you'll have to get another operative in the place of Mr Dibnah. The country is a bit hard up for men at the moment, especially men with skills, so he'll have to join up." I don't think Fred wanted to go into the army. He wanted to do more steeplejacking and I don't think it would have been long before he left my father anyway. At twenty-two I think he would have gone and done his own thing.'

Fred was glad he'd worked for Bill Rawlinson. He'd been there for six years, from the age of sixteen. Bill was a good craftsman, and he'd taught Fred a lot. One thing that was to stand him in good stead for the rest of his life was how to measure, as everything he went on to do, from making iron bands for chimneys to machining parts for his engines, involved measuring. But Fred couldn't avoid National Service. First of all he had to go for a medical examination. After that he was sent to see what Fred described as 'some sort of careers fellow'. Fred told him all the jobs he'd done and pleaded with him, 'Give me a job outside like building or engineering. If you keep me in the open air I'll do my National Service with pleasure.' But it was all to no avail and Fred was put into the cookhouse.

After six weeks' square bashing at Aldershot he was sent to Catterick to learn the basics of army catering, like how to fry a dozen eggs at once. Fred remembers there were forty men on the course and none of them had any experience of cooking. 'One,' he said, 'had worked a steam hammer; a few had been bricklayers and there were a couple of other joiners.' After his course Fred was posted to Germany and within six weeks he'd wangled his way out of the cookhouse. His regiment, the 14/20th King's Hussars, had bought an old farmhouse to stable their horses and kennel their foxhounds, but the place was a total wreck. Fred spotted an opportunity and let his commanding officer know about his property-repairing skills. He was entrusted with the job of restoring the roof and building brick stalls for the horses. His superiors were so impressed by Fred's work that he was employed as a jobbing builder and handyman round the stables and the mess and as a keeper of the horses and hounds. For Fred it was a perfect way to spend his National Service, especially when he found out that just by the stables there was a hole in the perimeter fence with a pub opposite.

The job wasn't too demanding and to pass the time Fred made his first weathercock out of an aluminium tray. He was so pleased with it that he put it on top of the stables. He was also interested in guns and he started to do drawings of them, which he put up on the wall by his bed. Fred had no complaints about his National Service. 'It wasn't a bad thing,' he always used to say. 'If National Service had been kept going it might have kept all the vandals and hooligans we have now in line.'

3

The Ups and Downs of Chimneys

Once his two years were up, Fred got out of the army. It was a great day when he boarded the plane that flew him back from Germany to Manchester. When he got home he spent most of the first few days drinking and then went across to the decorator's yard where he'd stored all his steeplejacking gear, to check that it was still OK. His ladders were outside in the yard, but they'd been covered up and no harm had come to them. The rest of the tackle was in a shed, all ready to be used. So Fred had some cards and bill heads printed and a few photographs of jobs he'd done before his army days.

Now that he was older and more experienced, he felt more confident about going into places and selling himself. Soon after getting back, he put on his best suit and set off round the mills of Bolton in search of work. A lot of the chimneys looked to be in quite a bad state, so he was sure he'd be able to get jobs without too much trouble. But the reality was different and tough times lay ahead. The Bolton of the early 1960s that Fred came back to was changing rapidly. These were the 'never had it so good' years of Prime Minister Harold Macmillan, but things weren't too good if you were a mill worker in Bolton. In the town no fewer than seventy mills totalling some 9 million square feet of production space closed between 1957 and 1965. The majority of these closures came in the early 1960s, when most of the town's long-established names disappeared. Among them were Vantona's Garfield and Gibraltar Mills, where Fred had met his engine-minder friend when he was younger. By 1967 only thirty-seven mills were operational and fifty others remained standing, ready to be put to alternative uses. It was against this background that Fred was setting up his steeplejacking business and it was to have a profound effect on the direction his business took.

Derek Roscoe had first met Fred while he was on leave from the army and he met him again soon after he came home to Bolton. 'As I rode about on my old 1927 AJS motorbike, I must have been coming past Burnden Park one day and there was Fred flagging me down. He must have seen me going backwards and forwards along the road over the years since I first met him. He told me that he was interested in buying my motorbike. He was mad keen on it, but I wasn't for selling. So we parted our ways. But after that we became firm friends and I went down there regularly to see how Fred was getting on with his work. He hadn't got a vehicle for carrying his stuff.

He had a pushbike, as I remember, and a trailer. By this time he had quite a few ladders for his steeplejacking and I suppose he was thinking of a motorbike to give him a bit more mechanical means of getting them around. But Fred was persistent and eventually he twisted my arm into selling the old AGS for twenty-one guineas. It had to be twenty-one guineas for Fred in his old-fashioned way. I'd have been quite happy with twenty-one pound but I got an extra twenty-one shillings which I weren't complaining about.'

But before Fred could ride the bike there was a problem he had to sort out. He got the bike insured and then he had to get it taxed, so he went down to the town hall with the log book and his insurance certificate. At this time Fred hadn't passed his test and it was only when he got to the town hall that he found out that, according to the law, if you were a learner driver you couldn't ride a motorbike with an engine capacity over 250cc. Well, his new bike was a 350. So they refused his licence and Fred was heartbroken. He went to see another friend, Alan Crompton, and told him what had happened.

'Fred, it's your lucky day,' Alan said. He'd owned a couple of other motorbikes and he just happened to have a log book for a 1935 250cc AJS. So Fred changed the registration plates on his bike, took the new log book down to the town hall, passed it off as belonging to his own bike and got a licence.

'He were made up! Thrilled to bits!' recalls Alan.

With his new motorbike, Fred had all he needed to get

his steeplejacking business on the road. But by this time a lot of the mills had already been shut down, and most of the others were struggling for survival. Their buildings, and especially the chimneys, needed repairing, but there was little or no money available. Fred found this out the hard way. To begin with he didn't think it best to ride a motorbike around the mills of Bolton, wearing his best suit, gold watch and chain, so he used public transport. But this meant that he could make far fewer calls each day. Then, when he got to the mill, they would keep him sitting in the foyer for an hour or more before telling him his services were not required.

Several weeks went by and Fred seemed to be getting nowhere fast, so he had to start looking for other jobs. For two or three days at a time he would do odd jobs for local property repairers: a bit of slating here or a new fall pipe there. From the money he earned he would give his mother something for his lodging, while the rest went on fags and the odd pint. While he was at his mother's house he converted the loft into a studio and broke into the existing chimney breast so he could have a fire up there. He built the fireplace out of stones from the river. To get access to his studio Fred cut a beautiful hole in the floor. It was three feet in diameter and so perfect, recalls Alan Crompton, that it looked as though Fred had drilled it out with a great big bit. The idea was that he was going to put in a spiral staircase. But that never happened; he made do with a builder's ladder. It was very low up in the loft but, because Fred wasn't very tall, it was OK for him. And that's where he did his drawings – mostly of steamrollers and guns.

Alan Crompton remembers Fred's little studio. 'I went up there with him and he showed me his drawings. The nicest one was a watercolour that he'd done at fifteen when he was at art school. The painting was of a builder's yard. There were no trees and no grass in the painting. It was all chimneys, wheelbarrows, bricks and builder's tackle.' In 1963 Alan moved into a little farm cottage adjacent to the Manchester–Bolton–Bury Canal. The cottage was in a bad state so Fred went over to do some repairs and maintenance work. He continued with little jobs like this for over six months, always doing three days' repair work and then going round the mills again for a couple of days. He was looking for real chimney work, but during that time he never got a single job.

At last, Fred had a stroke of luck. In those days there were three breweries in Bolton. One of them was near the art school he'd been to. The owner's house was next door, and Fred had been quite friendly with his son when they were lads. By now the son had grown up and was working in the family brewery. The brewery had a little chimney which had fallen into disrepair. In fact it was in a terrible condition. Partly because they were old acquaintances, but mostly because something had to be done about the chimney before it fell down, Fred was given the job. He spent three weeks mending it and making it perfect. For this he earned the princely sum of

£45. While he was doing the work he had no idea where the next job would come from.

The brewery stack needed a new metal band round the top, so Fred's father introduced him to a fellow who did some welding and metal-working. Fred talked to this fellow quite a bit as he was making the band. After finishing the brewery job the search began for the next one. Fred went to see the welder again on the off-chance he might know of something. 'It's a good thing you called,' the man said. 'I was talking to the vicar of Bolton the other day and he said he'd like to see you. Something about some work that needs doing on his weathervane.' This was exciting news because the parish church was the biggest building in town. Fred felt that, if he got the job, he'd be able to approach all the other vicars in the area. If they knew that he'd been allowed to have a go at the parish church, Fred felt, not unreasonably, that they would all let him work on their lesser churches.

Years later, when we were doing a programme on churches and cathedrals, I went to the church with Fred and he told me why it had such special significance for him. 'This large Bolton parish church tower launched me on my steeplejacking career about forty years ago. I were asked to gild the weathervane on top. I've done it again since – it's about eight years since I last stuck the gold leaf on that's shining in the sun now. When I first met him I thought the vicar here was a funny sort of fellow, who had a bit of a mad look about him. He drove a huge black 1929 Humber limousine. I'd arrived there on my 1927

AGS motorbike and the pair of us hit it off straight away. The vicar liked my bike, so from that time on I used it more and more often when I was going around looking for jobs.'

Eventually Fred managed to bring up the topic of work. 'Ah yes,' the vicar said, 'I want you to take a squint at my weathervane.' The tower of Bolton parish church was the tallest thing in the centre of town and from the ground the weathervane looked not much bigger than a postage stamp. In reality the main part of it was six and a half feet long and three feet deep, and it was made of extremely heavy metal. How those Victorian builders managed to get it up there in the first place it's impossible to know. It was going to be a bad enough job just repairing it. But Fred wasn't going to turn down an opportunity like this. Fortunately he didn't have to ladder the whole of the tower, as he was able to get most of the way up by means of a staircase on the inside. This came out level with four pinnacles, and from that point he was able to put up his ladders and then put scaffolding round them. It was worth the climb for the fantastic view from the top.

After Fred had fixed the weathervane the vicar said he wanted it gilded, which meant putting on paper-thin sheets of gold leaf, never the easiest of jobs at that sort of height on a windy day. But Fred got the job done and his picture appeared in the local paper, which was good for publicity. The vicar became a good friend and put a word round the other Bolton clergy, so that Fred got his next job mending the clock faces on another church tower.

But mending church steeples wasn't the same as working on chimneys, so one day Fred summoned up enough courage to visit the biggest firm in the area, Hick Hargreaves. It was quite a forbidding place with iron pillars, iron doors and an iron portcullis type of gate. It was a fine day and Fred decided to go on his motorbike, which looked quite impressive with its carbide lamps and a great bow horn. Fred roared into the forecourt and very nearly knocked over the boss. Fortunately the man wasn't too bothered because he was interested in the bike and wanted to know all about it. Next thing he was telling Fred that if he had a moment he'd take him round his works and show him all his chimneys. Right from this first meeting Fred got on very well with the boss and this led to regular work on Hick Hargreaves chimneys for many years to come.

The money that Fred made on his first job for Hick's enabled him to buy ten more ladders. This gave him thirty in all. Until then he'd done everything single-handed, which made the hard work all the harder, so he began to look for an assistant. He took on an old fellow called Percy Porter. Fred taught Percy how to tie knots and handle the equipment. Shortly after Percy joined him, Fred was summoned to a real giant of a chimney, 270 feet high. He had just enough ladders to get up it, but nowhere near enough scaffolding to go round the top. The engineer told him they wanted the top half taken off the stack. It was surplus to requirements and they had to pay heavy insurance premiums on it.

When Fred reached the top he found a beam across the centre for pulling up materials. It was made from a length of an old railway line. The old-timers who'd built the chimney had moved this beam higher and higher during construction. Once they'd reached the top they'd just left it there. Now Fred was faced with the task of getting it down again. The steel beam was too heavy to manhandle, so he set about cutting it in half with a hacksaw. The top of the rails was very tough, so he used seven blades just getting through it. Down at the bottom of the chimney there was a bit of waste ground and Percy warned everyone to keep clear of that area. The first piece that Fred threw down landed flat on the ground: the next one speared into the ground end on. After that it wasn't too bad, though he had to give up sitting on the rail as he sawed away towards the end.

Next Fred began taking down the chimney itself brick by brick. Apart from the hard graft, the operation was made all the more difficult by the smoke that was billowing out all the time. They couldn't stop production, and one way and another it took Fred six months to do the job. As he bashed away with his hammer and chisel, he used to think about all the men who'd gone before him and put up these great mills and chimneys. The chimneys had become so much a part of the skyline that no one took much notice of them any longer, but Fred had close-ups of beautiful stone- and brickwork up at the top. Another bonus for Fred was the view from the top of a two-hundred-foot chimney, especially during the winter.

There would be endless rows of terraced houses, all puffing out smoke to add to that of the big factory chimneys. The smoke would combine to form layers, through which the sun filtered, grey pink in the mornings and deep purple just before nightfall.

Fred was enjoying his work. Because much of it was on mill chimneys, he found out a lot about working life in the mills, usually from the engineers. But working in the mill didn't appeal to Fred. 'I don't really think I would ever have liked to work in one of these places. I actually experienced the noise of it all and it was deafening. I'm really more interested in the mechanics – the engines and the boilers that made it all go.'

There's one particular steeplejacking story that Alan Crompton has always remembered from that time. 'I'd been down at Fred's one early evening – it were just going dusk perhaps – and this young fellow appeared. He were a demolition man, and he'd got a contract to demolish a mill up Chorley Old Road – and the chimney. He couldn't do the chimney himself so he'd brought a steeplejack in, but the steeplejack had hit problems. So the demolition man offered the job to Fred.

'I took Fred up to see this chimney on my motorbike straight away and it were dark. We got inside the mill yard but the bottom ladder had been removed – quite rightly, so that children couldn't climb up these ladders. The ladders were about ten feet long, so I thought, how's Fred going to get up the chimney? As I was working this out Fred went round the back of the chimney and

reappeared with a floorboard. It was a very long floorboard, about eight or nine feet long. Fred leaned it on the
chimney and then like a monkey he crawled up the floorboard on his hands and knees and grabbed the bottom
rung of the ladders. After that he was up the chimney in
no time. It was dark by then but it were a moonlit night
and I could see him in silhouette, walking round the top
of this chimney. He was having a good look at it, trying
to decide whether to accept the job or not. I was down in
a backstreet watching all this when a gate opened and a
woman rushed out, shouting at Fred. She thought he were
a jumper!

' "Come down, come down!" she was saying. I don't
know whether Fred could hear or not. So I went up to her.

' "He's not a jumper, love, he's a steeplejack come to
survey the work," I said, but she was getting quite
agitated. So Fred came down to placate her. When she
knew what was happening this woman showed us some
of the damage done by the other steeplejack's amateurish
attempts. There were dinted and damaged doors in the
backstreet caused by falling bricks. But to cut a long
story short, Fred didn't want the job. It just didn't suit
him. But that were an experience, seeing him up that
chimney in the moonlight.'

Fred didn't just climb chimneys because it was his job.
He loved climbing and he was still up to his tricks and
practical jokes. Kearsley power station had three cooling
towers and Alan remembers how Fred got the idea that he
wanted to scale these overnight and paint *Oxo* on the side

of the towers facing Bolton in fifteen-foot-high letters. But he would need somebody at the bottom – an assistant pulling the bosun's chair round, so that he could make the letters appear magically overnight. And everybody would wonder . . . these massive letters! 'That was the sort of hare-brained scheme that Fred was capable of and it were intcresting to see how he scalcd the ladders. He were using just two short ladders about six feet long. He'd start off by leaning the first ladder on the base of the cooling tower and he'd climb up to the top rung with his rope and his plugging chisel and his lump hammer and he'd reach up and put a hook known as a dog in. Then he'd get the other ladder and hook it on the dog. He'd then get the first ladder, hang it on a hook on his belt, climb up the second ladder, reach up again as far as he could and put another dog in. And so he leapfroggcd up these great towers.'

Another time Alan went with Fred to look at an old dovecote. It was a cylindrical building about forty feet tall and it had been ransacked. The floors had all been burned out. Fred climbed up on to a wall, which gave him access to a slit window about two feet high by one foot wide. Fred, who was quite slim then, wriggled through the window and disappeared into the building. Alan followed him, got into the aperture and sat there on the window sill. By this time Fred was already on the opposite side of the building, where there was a spiral staircase. He walked up the staircase, but Alan remembers wondering how in heaven's name he got across there. 'There were no floors at all and when you looked down from the window, there were just the dirt foundations about eight feet below where the ground floor should have been. Fred then started to rub it in a little bit.

' "Come on up here, Alan," he said. "You can see all over Horwich. Come up and have a look at the view."

' "Oh aye," I said, "and how do I get across there, Fred?" He didn't tell me but I thought I'll wait till he's coming back and I'll watch how he does it. He came down the staircase and to this day I don't know how he did it. Fixed to the wall at floorboard height was a three-by-two spar. He stepped on to this spar but the wall was right up against his chest. How on earth he managed to walk on a three-by-two fixed to the wall I don't know, unless he had suckers on his chest. He was like a human fly. It's impossible to walk on something that's fixed to a wall. But he had the nerve and he'd got very strong fingers and strong arms. He was as slim as they come then and he could lift two fifty-six-pound weights above his head – one in each hand, strong as a lion. And he was mischievous in his own way. He just did dangerous things, but they didn't do anybody any harm. To me they were dangerous and to anybody looking on. But to him, I think it were just a test of his ability.'

They were the qualities that Fred needed for his steeplejacking business, which by this time was going from strength to strength. So Fred took on another man, called Joe Robinson. Joe was very strong, even though he was nearly seventy, and he proved to be a great help. One morning Fred arrived at a stack with Joe. There was a thick fog and they couldn't see very far up the chimney. 'You're not going up there to throw things down today?' said Joe, as if Fred was a bit crazy.

'Course I am,' Fred said. 'Work must go on. Just rope off a landing area in case any fool tries walking into the firing line.'

Once Fred got up to the top, there was an amazing change in the weather. It was like being in an aeroplane up above the clouds; the sun was shining and it was a lovely, bright day, which made Fred feel quite carefree. Dangerously carefree. He started chucking bricks down, but forgot that all the time he was doing this he was working his way round the chimney stack. Before long he was well away from the roped-off landing area. After a while Fred began to notice that when he threw a brick down, a different sort of noise came up through the fog. So he chucked another in the same direction and listened carefully. The unmistakable sound of breaking glass drifted up from somewhere in the gloom below. Next thing there were shouts. It was only then he remembered that, on the opposite side of the chimney from his landing area for the bricks, there was a very grand-looking office. It had a slate roof, big glass windows and a fancy decorative plaster ceiling. Below that was some very smart walnut panelling and what the company called their terrazzo hallway. But Fred was more worried about whether he'd killed anybody as he raced down the ladders a couple of rungs at a time.

When he reached the ground and looked towards the building, all the office staff were cowering in the doorway. They were covered in plaster, from the owner of the place down to the office junior. No one said a word to Fred as he wandered into the terrazzo hallway and up into their magnificent office – or what had once been their magnificent office. The first sight that met Fred's eyes was some thick cornice moulding lying beside the door. Beyond this, layer upon layer of roof plaster covered the beautiful terrazzo floor. Sticking out of the plaster were a lot of fair-sized chimney bricks. The desks, typewriters and even the teacups had all been given a good peppering. It was a scene of utter devastation and Fred feared the worst. But everyone was so relieved to have escaped without loss of life or limb that they treated it all as if it had been some sort of adventure. Fred was very happy to go along with that, and he came back to help them repair the roof.

After this Fred became very safety-conscious, not just for himself but for others as well. He kept a lookout for dangerous chimneys, and he developed a knack of being able to spot them a mile off. There was one stack that looked particularly precarious. It was above an old warehouse, where he had gone to buy a new cupboard for his mother. Years before there had been a colliery on the site, and the chimney stood over the unused engine house. Fred looked at the stack; it was in terrible shape. When he'd bought the cupboard, he found the warehouse owner and suggested putting his chimney right. But the owner wasn't convinced, even when Fred brought along his photo album to show he was genuine.

Fred tried to explain to him that the iron ring round the top was hardly holding itself together, let alone the chimney. 'Look at the way those bricks are bulging,' he said. 'There's a lot of pressure behind them. All you need is a bit of a change in the temperature and you'll have the whole lot crashing down.'

'You're just trying to scare me,' the fellow said, 'but you can't frighten me.'

'Well, be it on your own head,' Fred said. 'Don't say I didn't warn you.'

It was only a few days later when there was frantic knocking on Fred's front door. Standing outside was the man from the warehouse. He told Fred he had a problem with his chimney. Bits of it were beginning to fall off. As soon as Fred heard this, he was round to the warehouse quicker than the owner could get back there himself. He found the place already swarming with police and building inspectors. Something had fallen off the chimney and, as usual when anything fell off a chimney, everyone had panicked. The warehouseman was white-faced, but in a way he was a lucky man, because there was a house right next to the chimney and when the piece of chimney stack had fallen it had come down on the other side. If it had come down on the same side as the house, the falling masonry would have gone straight through both floors. There had been a little girl asleep in the top room and people downstairs.

The warehouse fellow changed his tune completely and asked Fred how much he would charge to do the job for him. Then he tried knocking Fred down to a rock-bottom price. But the job had to be done. Public safety was at stake. Fred took the top six feet off the chimney, put a new iron band round, and repointed ten feet downwards. Then, just before he was ready to remove the ladders, the warehouse fellow said to Fred, 'I've had some company name signs made; I want you to fix them up there at the top where everyone will be able to see them.' Fred tried to explain that you're not supposed to put signs up without planning permission, but the man replied, 'If you want me to pay you, you'll do as I say.' So Fred did, but he made sure that one of the signs was facing the town hall. As soon as he'd got it fixed, he whipped his ladders away. It wasn't long before the phone rang. It was the warehouse fellow again. The town hall had told him that the signs were illegal and that he'd have to get them down straight away.

'Right,' said Fred, 'but that will be another job.'

By this time there was a lady in Fred's life. His brother Graham remembers how they met. 'Our kid met Alison through her father at a pub on Manchester Road where they all used to drink. Alison's father was a collier and she came in one night to ask him something. And our kid said, "Who's that?" He said, "Oh, it's my daughter." Fred made the mistake of asking her father if he could go out with her, but her dad didn't like Fred. He didn't think he

Fred (*right*)
with his
brother Graham.

During his art
school days.

BELOW: The
Bolton skyline
in 1932.

TOP LEFT: *Beautiful Bolton.*

BACKGROUND: *A smoggy day in Bolton,* 1937–8.

Desolate Bolton street, 1937–8.

Fred, his father Frank, mother Betsy
and brother Graham.

National Service in Germany.

L. S. Lowry, *Going to the Match*, 1953.

ABOVE: Working on *Betsy*.

LEFT: Horwich Loco Works, 1960.

Top: The proud dad with Roger in his arms and Jack.

Centre: A pub stop on the way to a rally – *Betsy* with the living van in tow is last in the line.

Bottom: *Betsy* parked outside Fred's house before he built the extension.

CLOCKWISE FROM TOP LEFT: Fred on his motorbike, 1966; the young steeplejack; the view from the top of Bolton Parish Church, taken by Fred; Bolton Parish Church, 1962, Fred's first big break; Fred's first chimney, built on his mother's house.

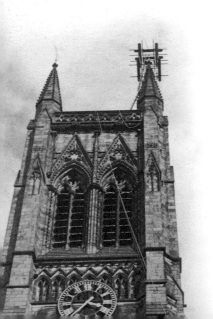

ABOVE: Ken Devine, Hallé Orchestra
lead violin, in steeplejacking mode
– taken by Fred.

RIGHT: Fred in his bosun's chair.

TOP: Monton, the end of a great chimney. There were no crowds at Fred's early fellings.

BOTTOM LEFT: Preparing for a felling.

BOTTOM RIGHT: After some anxious moments, the chimney at Cockermouth finally falls.

was good enough for his daughter and, being a man of few words, he just said, "Keep away, son." But when Fred saw Alison again he summoned up the courage to ask her if she fancied going out with him.

'Next thing,' remembers Graham, 'he was going out with her and within a few weeks they were engaged. I haven't a clue what her dad thought about it all at first! I was working on a school that was being built at the time and one day I saw my mother on the site, with my general foreman, and I thought, what's happening here? She was in tears. So I went to her and said, "What's up, Mother?"

' "Give me a handkerchief," she said. "It's our Fred. He's run away to Gretna Green, to get married."

' "What's wrong with that?" I said.

' "You know I wanted to put my new hat on and have a proper wedding," she replied.'

It was a dramatic decision that Fred and Alison had taken and Alan Crompton remembers some of the reasons for it. 'Apparently when he decided to marry Alison, there were that much argument over who was coming to the wedding and who wasn't, they said, right, nobody's coming, we'll elope to Gretna Green. Fred told me, "Her lot want so many coming and our lot want so many we've just had enough of all the fuss." As well as that, all Fred's pals who'd never thought he would wed anybody were pulling his leg, especially as Fred was ten years older than Alison. There was so much fuss that Fred suggested they should leave the lot of them and go away and get married quietly.'

By this time Derek Roscoe had left Bolton and moved up to Scotland to work as a chauffeur for a rich lady with an estate in Kirkcudbrightshire. 'Fred thought they'd be able to stay with Derek,' said Alan, 'so Alison left a note for her mother where she could find it and Fred left a note for his mother, where she didn't find it. I don't know if it were deliberate or not but his mother were going frantic till she found out that he'd eloped.' The runaways had caught a train from Bolton to Carlisle and then a series of local buses to Dumfries. The trouble was, because it had all been done in such a hurry, Fred hadn't let Derek know he was coming and, worse still, he didn't have his address. So Fred and Alison went to Dumfries bus station and asked all the drivers if they knew Derek from Bolton and where he worked. Unbelievably, when they got to the last bus in the line the driver said; 'Oh aye, I know the Sassenach chauffeur. Get on and I'll take you there.'

Fred and Alison were dropped off at the end of a track leading up the mountainside through a pine forest. They set off through the mist and rain. It was all sheep and cattle and hills, not the sort of landscape that Fred was at home in. By this time it was pouring down and when they finally got to the house where Derek was living, he wasn't there. He'd gone back to Bolton on his holidays. So they were stuck and by this time it was getting late in the day. Eventually they found their way to a farm a bit further up the track, and when they told their tale they were invited to spend the night there. Derek only found out when he got back from his holiday. 'And ever since,' he said, 'that farmer has been fair proud that Fred stayed.' Fred slept with the farmer, who wore an old nightcap with a bobble on the end, and Alison slept in a wooden cot that was set into the wall.

Next morning the rain had stopped. They got a lift down the mountain and waited for the bus at the place where they had got off the day before. When they got to Gretna there was another setback. They found out that you had to be resident there for twenty-one days before you could get married. This was a serious blow for Fred. His money was running out fast and they would have to find somewhere to stay for three weeks. But, resourceful as ever, Fred noticed that the gable ends of one of the hotels needed repointing, and he did a deal with the owner to do the work in exchange for their bed and board.

Alison wanted a church wedding so they married in the church at Gretna Green. Derek Roscoe recalled the re-action back in Bolton. 'Really they were both persona non grata about this whole affair because they never had a proper wedding. But it suited Fred because it were dramatic and he was able to be a bit of a showman again, which is what he liked. Anyhow, good old Fred; that girl loved him. Alison loved him for himself; she really did. And I was pleased he'd got married. There were a time when people said, "Is he or isn't he?" – regarding girls, because he never seemed to talk about girls. He never looked at them. Women were a word that would never come into his vocabulary at all. So I were quite relieved when he got Alison and she were a cracker. She were a grand lass.'

In spite of the upset it had caused in the family, the newly-weds went to live with Fred's parents. That is never easy for anybody, yet somehow they all made it work. But naturally Alison and Fred wanted a place of their own. They looked at one place after another, but there was always a problem. Then one day the Earl of Bradford's agent, Stanley Boardman, rang asking Fred to come over to his office. Fred thought it was going to be another job until the agent said, 'We have a bit of property that might interest you and Mrs Dibnah, as a family house, if you take my meaning.'

'What's it like and where is it?' Fred asked.

'It's a Victorian gatehouse,' the agent replied, 'and it's just off Radcliffe Road.'

'That's a bit of a posh district,' said Fred.

'Well, the place needs doing up a bit,' the agent admitted, 'and I thought what with your capabilities . . .'

'We'll have a look at it then,' Fred said.

Mr Boardman took them there himself. It was a fair-sized place, which Fred and Alison wanted, but he'd been right about the repairs. The old gatehouse was beginning to stick out like a sore thumb in that smart residential district. But it was a fine-looking house and Fred knew that with a bit of time he'd be able to do all the work that was needed himself, so they decided to take it.

* * *

As they settled down in their new house Alison had to learn to live with Fred's steeplejacking activities. At first she felt the less she knew the better, so she tended to stay away from any chimney that Fred had his ladders on. In June 1968 Alison and Fred had their first child, a daughter, whom they christened Jayne. But Alison was anxious about Fred's work so he persuaded her that she might be reassured if she came to see him at work. She wheeled Jayne to the job in her pushchair, but seeing Fred up there in the sky made matters worse. Eventually she accepted that Fred was a professional who always kept within safety margins, so she learned not to worry too much.

When I talked to Fred later he used to say, 'The everyday steeplejacking was pretty safe and I was never worried about falling off. There was always an instinct for survival because you knew at the back of your mind that if you put one foot wrong it would mean a visit to the undertaker, but if I'd worried all the time about what would happen to the wife and kids I'd have been a nervous wreck.'

Fred was a happy man. His reputation as a steeplejack was growing and he was getting plenty of work. He had a wife who could help to keep him organized and Fred said, 'She always looked good, especially when she was dressed up.' Above all, they'd got their own place. The house needed to be repaired and redecorated and all this work had to be done at times when Fred didn't have any other jobs on, which meant he was always short of money when he was doing it. The winter months in particular were hard, with jobs difficult to come by. During times like this Fred discovered how lucky he was to have a wife willing to take the rough with the smooth. Fred had his own strange way of rewarding Alison, by letting her light the fires under chimneys that had to come down. It became a sort of tradition. Alison also helped with the books, and even with debt-collecting.

Debt-collecting was an important part of the job. The chimney repair and demolition business was a peculiar one. Fred often had to deal with so-called property developers, who would buy up old mills and rent them out to other businesses. Usually no one gave much thought to the chimney stack still stuck up there by the side of the mill; or at least that would be the case until it began to fall apart and became a liability. There was one particular chimney beside an old mill that belonged to one of these property developers. Fred had heard that none of the other local steeplejacks would go near the stack, let alone climb up it. The chimney was a nasty one, and everybody said that the man who took on the job would end up dead. On the other hand, the offer being made by the property developer kept going up. So there was a waiting game going on, with the chimney crumbling, the other jacks saying no, and the developer all the time raising the fee. Fred was interested. Apart from the money, which he desperately needed, the stack presented a challenge. Demolishing it would mean his reputation would rise along with his bank balance. But it was risky. He talked it over with Alison, and she agreed that he ought to do it.

The first thing Fred had to do was to negotiate with the property developer. The man turned out to be very courteous and he listed his conditions in the clearest possible terms. If Fred managed to fell the chimney without damaging any of his property, let alone maiming or killing anybody, he would pay him the agreed sum within one month. Fred insisted on getting this in writing and the developer agreed.

Fred did a perfect job. Even the developer who came to inspect it agreed that it couldn't have been bettered and he assured Fred that he would pay him within the month. But getting the money out of him proved impossible and fifteen months later Fred still hadn't been paid. Generally Fred was able to look after himself, including getting whatever money was owed to him. But this time he couldn't get it, whatever tactics he tried. Whenever he went to the developer's office, the girl at the reception desk would look straight through him and suggest that he should ring or write just like other people did. But when Fred did write, his letters went unanswered, and when he phoned, he only ever got some underling, who said he would look into it and ring him back. But he never did. Fred kept trying in vain to get through directly to the developer, but he was either at a meeting or off doing a deal in another part of the country. His secretaries called them conferences and he seemed to have hundreds of them. But, whatever tactic Fred tried, the cheque never came.

Fred really needed the money. Christmas was coming and Alison was talking about extra expenses and presents. 'Eventually,' Fred recalled, 'Alison suggested that she should go and collect the money. She went off one morning after breakfast and, halfway through the afternoon, she came back with the cheque. It was just in time for me to get it into the bank.' When he got back Alison told him how she'd succeeded where he had failed. To start with, she wouldn't be put off by the dolly-bird at the front desk. In fact, the young woman was sent upstairs with a message that Mrs Dibnah intended to sit in the hall for as long as it would take before the money that had been owing for fifteen months was paid. People came down from the accounts department and tried to put her off; then they tried to reason with her; and then they left her waiting through their dinner hour. But Mrs Dibnah wasn't going to move. Finally, after trying the 'it will be in the post' trick, one of the directors of the company brought the signed cheque down to her.

Problems like this with payments were not uncommon, but steeplejacking was a bug and Fred had certainly been bitten. Then somebody else appeared on the scene who'd got the bug – quite an unlikely character in many ways. It was while Fred was working on a warehouse chimney that he'd first noticed a young man hanging around. Fred got talking to him and found out that this chap was a violinist. But he wasn't just any old violinist. His name was Kenneth and he played in the Hallé Orchestra; in the front row, so he must have been good. After the warehouse job Fred had to take down a massive two-hundred-foot chimney for Courtaulds. It was going to be a big task, and when he heard about it Ken asked if he could give a hand. So, instead of practising for his next concert at the Free Trade Hall in Manchester, he came and helped Fred ladder the chimney. Ken liked the feeling of being on top of the world just as much as Fred did and, with his help, Fred managed to knock the top fifty feet off the chimney.

Then things began to get a bit difficult. Some of the bricks were hard and others soft, and the same with the mortar. It meant that nothing would break away easily. It seemed those old chimney builders must have known a trick or two that Fred didn't know. Ken was as keen as Fred to find a way of cracking the job, and he began to neglect his violin practice. One of the things they thought of was to take a road drill up there and have a go with that. It was a bit precarious, and all to no avail because that didn't work either. Another problem was that Fred hadn't quoted for the job, and, as the weeks went by, it was becoming a bit expensive. Ken and Fred racked their brains. How were they going to get this thing down? They thought about trying a few poppers, but Courtaulds didn't want them to use explosives. They devised some incredible mechanical contraptions and tried a fifty-ton hydraulic jack, but still the chimney held together. Then, all of a sudden, it split right down the middle. One piece, about four feet square, went down the centre; another that was twice as big bounced off the side and went through the roof of the blacksmith's shop, hitting the main shaft that drove all their machinery. Fred turned to Kenneth and said, 'Well, at least you have a job with the Hallé. I think my steeplejacking career has come to an end.'

It was dinnertime when this happened and the chief engineer had been playing billiards in the canteen at the other end of the works. One of the lads went racing across to break the news. 'Fred's just wrecked the blacksmith's shop.' Luckily for Fred, the chief engineer was a Sir Francis Drake type.

'Anyone killed?' he asked, still squinting along his cue.

'No, sir.'

'Oh well, I'd best finish off this game.'

But the chimney itself still wasn't finished and it put up a stout resistance right to the end. When the fifty-ton jack didn't do the trick, Fred resorted to even more desperate measures. He brought in a steamroller and pulled off lumps of masonry with a wire halter. When it was over, Ken and Fred went to a pub and got themselves inebriated.

In spite of the odd difficult job like this, Fred was enjoying his steeplejacking work and so was Ken. Fred always described him as a frustrated steeplejack. If the slightest opportunity presented itself, Ken would be off to to climb a chimney with Fred in spite of the fact that his wife, Ann, was always asking him to stay off them. On one occasion Ken had been up on a chimney all day with

Fred and he'd got a concert that night. They'd stayed out too long and time was pressing, so Fred volunteered to give Ken a lift to the concert in his Land-Rover. Just as they were setting out the phone rang. It was Martin Milner, the leader of the Hallé Orchestra. His car had broken down and he was stuck. Could Ken give him a lift? So Fred and Ken made a detour to pick Martin up. The three of them crammed into the two front seats of the Land-Rover and headed for Belle Vue, where the concert was to be held in the King's Hall. As they reached the gates of Belle Vue, they had to join a queue of smart cars with even smarter occupants. When they got to the commissionaire on the gate he said, 'You lot can't come in here in that.'

'In that case,' said Ken, 'you can turn all this lot away, because without us there won't be any concert tonight.'

Although Ken tried to get to as many chimneys as he could with Fred, he was what might be described as an enthusiastic amateur. He couldn't give them his full attention, on account of what Fred called 'all the Brahms and Beethoven stuff', so he introduced Fred to his uncle John. John was one of those people who seem to have done everything in life, including playing rugby, working as a coal miner and sawing stone blocks. Just the chap, Fred decided, to be his assistant. Their first job together involved demolishing a chimney for a firm at Croston, a little village on the way to Southport. The chimney was in an awkward position, between the works and a river. The river had only just been cleaned out and the local water authority told Fred that he must not drop a single brick in it. Anyway, John and Fred went through what had become Fred's usual routine for felling a chimney. First of all they cut away some of the base, put in wooden props and laid a fire round them. Tension mounted when they lit the fire and the props were burned away, but the chimney went over just right.

Not long after this, Fred got a council contract to dismantle a couple of chimneys. The first was within twenty feet of a main railway line, so the contract stipulated that it had to be taken down brick by brick. Fred went to look at the chimney and saw that it leaned away from the railway. It looked easy and, as he was confident of his abilities after so many recent successes, he decided to ignore what it said in the contract and do it the easy way. After all, the other way would have meant hammers and chisels for a week, and once it was done there wouldn't be anything the council could do about it. Fred chose a bank holiday Saturday to have a go at the chimney. There weren't a lot of people about when they started the job. When they set fire to the props they burned away but the fire went out with the chimney still standing beside the railway line. By this time it had gone dark and they had to leave for the night.

That night Fred rang Ken, his father and one or two other fellows he knew. He asked them if they could make an early start next morning. By 6 a.m. they were onsite, and within half an hour they had built the fire up again. This time the blaze took a good hold as soon as it was lit and within minutes thick black smoke started to billow from the top of the chimney as the timber and tyres that were packed into the base of it ignited. The early morning sky turned black with smoke, and soot showered down as the fire roared up the chimney. As the fire grew in intensity the pit props holding up the chimney began to buckle and collapse. Tension built; the railway line was very close, but Fred was confident it would come down in the right place. After about twenty-five minutes a huge crack opened up in the back of the chimney and within seconds the huge stack began to topple in exactly the direction that Fred had planned for it. By 8.30 a.m. the remains of the chimney were lying on the ground, without a brick anywhere near the railway. It was a perfect job. Time to repair to the pub.

A couple of days later Fred received a letter from the council. As he opened the envelope he was thinking that this must be his cheque. Very prompt payers, he thought. But inside there was a letter. Dear Mr Dibnah, it began, your actions were in breach of your contract, and also put us in a most difficult position with the railway. Since you have broken the contract it is now null and void. They also said Fred couldn't expect to be paid for the first chimney, or even allowed near the second, in spite of the fact that it was surrounded by fields on all sides.

Fred decided to call a few influential people he knew to get their advice. What each one of them told him amounted to the same thing. 'Creep, creep like mad,' they all said. 'Write a letter to the borough engineer and ask for an interview.' Fred did as he was advised and the interview was granted. The man responsible for the council's demolition was there, determined to shoot down any of Fred's arguments. Fred knew he was in the wrong. The contract had been broken, but Fred decided the best form of defence was to go off at a tangent, so he pointed out that many other chimneys felled in the town, by the explosive boys, had gone wrong. Fred pressed on: his method might seem old-fashioned, but the success rate proved its worth.

'I'll tell you what,' Fred said. 'I'll bring your second chimney down free of charge, if you'll pay me for the first one, which is already safely down,' and to the borough engineer he added, 'and I'll let you light the fire yourself.'

The engineer fancied this idea, and they agreed a date for the felling. When the time came the weather had turned nasty and there was a force-ten gale blowing. Normally Fred would have put the firing off and left the chimney on the sticks, but the big man from the town hall was still coming, with all his entourage. As well as this, the local newspaper was there to record the event. So the borough engineer lit the fire, and the newspaper photographers took pictures of him doing it. Then things started to go wrong. Besides burning the props, the high wind set fire to nearby grass and bushes. The council men and the rest of the spectators retreated some two hundred yards, leaving John and Fred gazing up at the chimney. The props had long gone from one side and the other had been weakened to ensure a fall, yet the stack stayed where it was – or rather, it swayed some three feet this way, then the other. Something had to be done. So Fred and John dragged their fifty-ton hydraulic jack out of the truck and got in dangerously close to the chimney to apply some extra pressure. Luckily they succeeded and the chimney came down, to a cheer from the distant spectators. When Fred went to talk to the council officials afterwards, it was all about fine balancing and counter-acting high wind forces. After all, it had to sound as if he had known what he was doing.

In spite of the fact that Fred was getting plenty of chimney-felling work it was, strangely enough, the job he liked least. His preference was for restoration and he would always point proudly to chimneys and church steeples that he had repaired. But as the mills closed down and the demolition men moved in, the chimneys had to come down with them. And sure enough his next job was another felling. It was a huge chimney at a big spinning and weaving company not far from home. This great stack was two hundred feet high, octagonal in shape with a square base, and it had beautiful moulding round everything. It was also six feet thick, and hammering holes through six feet of perfectly laid Victorian brick-work could be a difficult and somewhat painful experience. But they had to stick at it because once you've accepted a job, it's yours.

The job after that was a tricky one, another two-hundred-footer, which had to fall along a sixty-foot-wide slot with a street on one side and the wall of a mill on the other. John and Fred did all the cutting work very carefully. The chimney was due to come down on a Sunday morning, when there would be less traffic around. On the Saturday night they checked everything before leaving work. It all looked right: that huge stack was ready to come down. When Fred got back home the television was on and while he was talking to Alison in the kitchen he heard the newsreader say, 'And now here's a chimney that went wrong.' Fred raced into the front room, just in time to see a chimney come crashing down on a mill. When the newsreader said the site was in Yorkshire, the relief was terrific.

Sunday morning dawned and Fred was up early. When he got to the chimney the police were there and the usual sightseers had started to gather round waiting for the free show. Fred took a last look at everything, then as he was lighting the fire he said to John, 'Lancashire has got to do better today than Yorkshire did yesterday.' The flames rose, the props burned away, and the stack fell in a dead straight line, right along the yard. A couple of factory windows were broken by bouncing bricks, but they let Fred off for that.

Once a chimney had been felled, Fred always went straight off to the pub with the lads. Eddy Chatwood was one of them. 'You could never get Fred out of the pub. If you shouted, "Last chance of a taxi," he'd always mishear you and shout back, "Aye, I'll have another Guinness." That was a regular one. It was difficult to get him home at times. He certainly liked his Guinness.'

On the subject of drinking and steeplejacking Fred said, 'With most of the steeplejacks I know, people would say, "He was always drunk when he went up!" I wouldn't say that I've ever done it drunk, but if you were banging away with a big hammer all day, a few pints at dinnertime never did any harm, it sort of kills the pain! There's no doubt about that. I've had characters working for me who've got the shakes and have a hell of a job climbing up first thing in the morning. But after they'd had about five pints at dinnertime, they were full of beans!

'In twenty years I only ever had two calamities. One was the place with the terrazzo floor when I started chucking the bricks down on to the roof through the fog. The other was when the bricks fell on top of the black-smith's shop and completely demolished it. When that happened I thought it was the end of my steeplejacking career. I can laugh now but I didn't then. When I came down and went over to the pub for a pint I was shaking all over. But Mr Courtauld decided that what I'd wrecked wasn't worth keeping anyway, so there was only a very small claim off the insurance people. I had to have a third-party insurance policy for big companies like Courtaulds. If I killed one of their operatives, I'd be in dire trouble! They'd commandeer my steamroller or something like that. Insurance men and me don't mix.'

The chimney-felling clearly involved more dangers for other people and their property than the straightforward steeplejacking side of the business. And it was this sort of work that Fred was getting known for. Tom McFarlane, the steeplejack Fred used to watch when he was younger, said, 'I didn't get involved in felling chimneys the way Fred did, because my insurers wouldn't entertain felling them. They said it was a bit too hazardous and they could always go any way without you being in full control. I think Fred took some chances, but basically he always did a good job.'

In Fred's case, though, he'd weigh up the odds and, in spite of the risks he appeared to take, he was very good and wouldn't take chances that would endanger his life or anybody else's. 'The thing is that I don't intend dying,' he always used to say. 'It's a bit like being a motor car racer, or this fellow who jumps over cliffs on his motorbike. I bet he's not got a life insurance policy. If he has, it'll cost him a fortune. I don't really bother with that side of it. I know one thing though: I'm not going to die. There's no doubt about that. I'm going to die in bed with my boots on!'

4

All Steamed Up

Steeplejacking was now Fred's job, but his passion was steam and, just like his love for chimneys, it went back to his childhood. 'My earliest memories, on dirty windswept nights in winter, were the sound of the steam engine whistle and the main-line locomotive as it approached Bolton.' From his bedroom window Fred could see the greeny-yellow glow of the signal box down a narrow entry in between his uncle Fred's temperance bar and the local barber's. The locomotive would bash across the end of this entry and, as it did, Fred would catch a fleeting glimpse of two characters crouched in position on the footplate. The firehole door would open and he'd see the driver and the fireman silhouetted against a shaft of light coming out of the back of the cab. When it had flashed past, the scene reverted to the peaceful, calm light that had gone before. 'What an unbelievable feeling of power and excitement it must have been to stand on the footplate on one of them things. That's really what inspired me,' said Fred.

All day long, around Burnden Park, there were the sights and sounds of the railway at work. Everywhere you went you heard a locomotive clanging away. Near the end of Fred's street was a huge marshalling yard, and Fred was fascinated by it. 'There used to be a saddletank locomotive there with millions of wagons. As kids we all used to call it Fat Nancy. It were a wonderful thing that squealed when it went round bends; that wonderful squealing noise that steam locomotives made in them days that you never hear now. Today people rant and rave about noise pollution and anybody making a noise after seven o'clock at night. Yet all night long when I was a kid we had the noise of the shunting engines, terrific bloody noises!' The engines were used not just for passenger services but also to meet the demand for coal from the hungry cotton mills.

Not far from the first house Fred lived in, on Crescent Road, were the engine sheds, 26C Bolton Shed. It was only a slight detour on Fred's way home from school to go to the sheds. 'I used to look through the gate and occasionally got brave and ventured in. The vision in there never left me. There'd be row upon row of steam locomotives with fires in them all getting steam up and in the winter, when it was nearly dark, you could look through the smoke and all the windows of the little offices at the back of the sheds were yellow in the corners. And then there was always that wonderful smell and the grime and the dirt and fog; and there used to be coal and dust and black oil everywhere. To me it were quite romantic. But what a modern-day environmentalist would make of it, I don't know.'

Graham Dibnah used to go to the sheds with Fred. 'The train sheds were only about five or six hundred yards from where we lived and there were four or five of us who were all into trainspotting. We used to go up to the sheds nearly every day, to see what we used to call "namers". And we had an uncle who worked on engines, and he used to let our kid ride on them, and put a bit of coal in the fire.'

Fred never forgot that man who gave him his first ride on the footplate. 'All small boys have somebody they call their uncle, who isn't really their uncle,' he said, 'and I had one of them when I were a little lad. He drove an Aspinall locomotive on the marshalling yards near Trinity Street Station. On Saturday mornings I used to ride with him. It were quite fantastic really because the thing only had half a cab, and when it were raining hard, you had this wonderful piece of canvas that went from the edge of the cab down to the tender and with the fire-hole door open it were quite snug and warm; a bit like being in a tent with a big fire in front of you. He used to give me a drink out of his brew can which he had on a table over the top of the firehole door. But he weren't the only one. A lot of my relations worked on the railway. There were one or two engine drivers and one or two signalmen, and some did the work up the telegraph poles by the side of the track. Maybe that's where I got my climbing ability from.'

When Fred was a child, the best present he ever got was a model steam engine. It was about eighteen inches tall and ran on methylated spirits. It had a proper piston and slide valve, a pressure gauge that worked just like a real one and a shiny brass boiler. Fred loved this little engine and would play with it for hours on end. One Sunday evening he was playing with it in the parlour on a card table. The engine was whizzing round and, in a vain attempt to get more power out of it, Fred wedged a die-cast Dinky toy car on to the safety valves. This had the effect of holding the dead-weight safety valve down. 'The whole thing was going like a bat out of hell,' Fred recalled. 'All the table was shaking. Mother and Father were both in the kitchen listening to Albert Sandler's Palm Court Orchestra. It was half past eight. Then there was a bloody great bang. The Dinky toy blasted off up into the sky and stuck in the parlour ceiling, followed by a big plume of boiling water. The water started to drip off the parlour ceiling on to the couch and the best shiny table in the middle of the parlour. I didn't have long to wait before Mother arrived, and I got a great bollocking for mucking up the ceiling. We had a patch on there for years after.' That was Fred's introduction to being a steam-engine owner.

About the same time he started to see the real thing. He used to go to the fairground on Moor Lane in the early 1940s with his mother and father. When he was there he saw a great steam engine that had wheels seven feet in diameter. It was the only one left on the fair and it looked magnificent as it stood out in a dreary line of diesel engines. Fred was fascinated by it and every year he kept going back to that fair to have a look at it. Then one year it had gone, leaving just the diesels. 'Steam engine?' said a fairground man. 'Oh, they're obsolete now, scrapped. Yon diesels are replacing them all.'

As Fred got older he kept his interest in steam engines and in railways in particular. 'When I started going to the pub for the first time, at fifteen or sixteen, the pubs near our house were always full of engine drivers. I used to talk to them and then I got lucky. I met one driver who immediately recognized my great interest in engines and locomotives and the like. "If you want to ride on my engine," he said, "buy a penny platform ticket and go and sit at the end of the platform on Trinity Street Station tomorrow night." I sat there the next night just as the driver had said. It was a dirty winter's night and this great thundering engine came rolling in and stopped dead level with me at the end of the platform. After a quick look up and down to see if there were anybody of any authority about, the driver gave me a wave to jump up on to the engine. Then we steamed about twenty miles through the night. Back then in the days of gas lamps it were quite exciting. It were near the end of the steam era and the fireman knew there were no future for him so he didn't give a monkey's and he gave me the job of firing the locomotive. I remember him saying to me, "It doesn't

matter what you do, as long as you don't lose the shovel."
It were really good fun and I finished up doing it lots and
lots of times.'

On Manchester Road near Burnden Park there was a
pub called the Wagon and Horses, one of those Fred used
to go into to talk to the engine drivers. At the back of the
pub there was a yard you had to cross to get to the Gents.
Behind this yard were the railway sidings and on the back
wall there was an iron ladder down from the sidings. The
ladder was worn and shiny from constant usage as the
railwaymen came down to the pub for a pint. Fred always
remembered the scene in that pub. 'You couldn't actually
move in the vault for engine drivers and shunters and all
sorts of different railway operatives. Many a time there
would be a Black Five parked on the bridge with forty
wagons of coal heading for Yorkshire. I could never
understand why they brought coal trains from Yorkshire
into Lancashire and went back from Lancashire into
Yorkshire with wagons full of coal; but somehow that's
what they used to do. Anyway, these two mad Yorkshire-
men would be in the pub with their engine on the bridge
parked over Manchester Road with all these wagons on.
"We'll just have another before we go," they'd say, and
out there at the back of the pub the engine would be
blowing its head off with the safety valves roaring in the
sky. Then, after another pint, up to the top of the ladder
they'd go and get back on to the engine, and with a blast
on the hooter they'd set off back to Yorkshire. Quite wild
days of engine-driving they were.'

Fred was in the Wagon and Horses one Friday evening when a railwayman came in complete with his lamp and his hat and all the rest of his gear. The lamp had *Lancashire and Yorkshire, Manchester Victoria* stamped on the side of it. It turned out that the man had retired that day, which was why he had all his gear with him. Fred asked him what he was going to do with the lamp. 'Buy me a pint,' he said, 'and you can have it.' So Fred bought him a pint of bitter for one and fourpence (7p) and acquired a railway lamp that he kept for the rest of his days.

When Fred was a lad, steam locomotives were not just operating on Britain's railways; some were still being built. They operated on the railways right up to the 1960s and lots of big mills in Lancashire were still driven by great big reciprocating steam engines until that date. John Gourley was a boiler inspector who first went to Bolton in 1958 to inspect the engines and boilers in the mills. 'It was mainly Victorian vintage stuff,' he recalled. 'Most of the boilers I examined were over seventy years old, and many were much older. A few that I looked after were a hundred years old. They were coming to the end of their days, and by that time a lot of them had been repaired. I don't think people appreciated that the original workmanship had done remarkably well to last that long. And a more modern type was coming in then, which required far less maintenance. So I suppose people looked down on the Victorian engines.'

When Fred was working as a joiner, he used to sneak

into mill yards and climb up the engine-house steps to look at the engines. They were very impressive – a big mill engine might have a thirty-five-foot flywheel and half-inch ropes going to wheels up five storeys of spinning mill. When Fred started work as a steeplejack his job would take him round all the mills and factories. He would go into the boiler houses and engine houses and look at the beauty of the Victorian workmanship.

The steeplejack Tom McFarlane used to look at them as well. 'At the time I were steeplejacking the engine houses were spotless, really spotless. They used to always say you could eat your meals off the floor, and you could, it was that clean. And the engineers in charge were just like sergeant majors, really strict, but they had to be. If that engine stopped, the mill stopped, so it had to be spot on all the time.'

By now Fred had got his motorbike and that was when he started doing a bit of local touring, going to one or two of the early steam rallies with his mate Alan Crompton. Alan remembers some of those trips. 'I think the first adventure we had together on our motorbikes was when we went to Sandbach in Cheshire to a steam and vintage rally. I had a sidecar and my wife, Kath, came with us. Fred rode his AGS and we had a good day at the rally looking at all the vintage vehicles. He hadn't got a steam engine then but Fred was soaking it all up. When the time came to come home we always had to stop off for a pint somewhere. So we stopped in Lymm village and had a pint there. Both of our bikes had acetylene lamps that needed water and matches to make them work, and things started to go wrong after we left the pub because we ran out of matches and water. We were coming back on the East Lancs Road and it was very dark because it had no lights back in them days. I thought we'd missed the turning for Worsley, so we humped the bikes across the central reservation and were just setting off in what we thought was the right direction when a police car came and stopped us. They'd seen two bikes with no lights on and I thought, we're in for it now, we're in trouble. But Fred didn't seem bothered and he said brightly, "I'm sorry, Officer" – he always called them officer – "but I ran out of matches and water." They didn't know what he was talking about until they walked over to the bikes and saw they were vintage motorcycles. So Fred got in the police car, leaving me and my wife there, and they took him to Boothstown Rescue Station – a miners' rescue station. They gave him a box of matches and a bottle of water and brought him back. Fred could talk his way out of anything, because he had charm. He was such a likeable guy. He were so friendly, he got away with things like that with a smile.'

On these trips they would stop off at some of the coal mines in that part of Lancashire to have a look at the steam winding engines. Fred told me about the noise they made. 'If you had a colliery winding engine winding in a shaft that was a hundred yards deep with a cage hanging on the end of a rope with maybe ten tons of coal in it and the engine had to make a standing start, there would be a terrific noise. In the middle of a place called Leigh there was a pit called Parsonage. There was a winding engine there that you could hear in the next town down the road when it set off with a great cloud of steam over the top of the engine house. What a sight it was. And the roar that it made was like twenty locomotives going up a big hill with a lot of coal behind them.'

On another trip with Fred, Alan remembers they went to a rally at Harewood House near Leeds. Fred had his best suit on because he'd got no motorcycling gear. He had his cap on, but no goggles, no gloves and certainly no crash helmet. As they were on their way over the Pennines it started to rain and when they got to Harewood House Fred was wet through. 'But,' says Alan, 'Fred was tough. I said to him, "Fred, what are you gonna do?" He said, "It'll dry." That were Fred. And so it did dry. Anyway, we did the rounds at this rally and then the finale involved all the vehicles going into the arena – steam engines, vintage cars and vintage motorbikes. And so we all went in. I did one lap and came out. Fred didn't. He just kept going round and round because he loved showing off his bike. It were pure showmanship. He did about five or six laps. I think he were the only one left in the arena. He always loved to show off.'

By this time Fred had really got the steam bug and he wanted a steam engine of his own. He started to look round and let people know he was looking for one. He was soon contacted by a farmer he knew who owned a traction engine. Gerald Adamson told Fred that he knew another farmer who had lost one of his cows in an old army camp near Warrington. When he went looking for his cow he happened to look through the window of an old hangar and saw a steamroller. He told his friend about his discovery because they had a mutual interest in steam and then Adamson got in touch with Fred. The roller belonged to two Welshmen who dealt in second-hand military vehicles. They'd bought it off Flintshire County Council, probably for about £35, and they'd taken it to the camp and tucked it away in the hangar along with some army vehicles. Fred went to Warrington to look at it with his farming friend. 'We climbed into the camp over the railings,' he said, 'really breaking and entering in a way, and we got into this building where the roller was stored and it weren't very secure. We weighed it up and it was evidently fairly old, turn of the century it looked like, and it were very, very rusty with millions of pan washers on the tender to stop the water running out. There was an office there so we went in to try and find out who owned it, but there was nothing in it other than a few bits of paper. We found a piece that had *Croup Brothers – Second Hand Commercial Vehicles* written on it.' Fred went home and wrote a letter to them but he didn't get a reply.

All this happened before Fred married Alison. 'I was doing a bit of courting at the time,' said Fred, 'with this bird who had a Triumph Spitfire and I remember her picking me up in it – me with me waistcoat and me gold watch chain like John Steed, and her like Emma Peel –

and I went over there with her. We got there and bloody hell, what a mess it was, just like a junkyard with a right scrappy-looking house. I knocked on the door and when it opened there was a guy standing there, and I said, "I'm the man who sent you a letter about that steamroller that's at the army camp near Warrington." He said, "I'm sorry I never answered but the wife deals with all the letters and she's been ill. Anyway," he said, "give me £175 and you can have it." ' But £175 was a lot of money for Fred back in the 1960s and it wasn't something he could rush into. 'Hang on a minute,' he said. 'I'd like to see what the boiler's like and we'd like access to it so we can have a hydraulic test done on it.'

As chance would have it, the farmer's brother worked for an insurance company and he said he'd get a boiler inspector to take a look. Fred came back with the boiler inspector and gave the engine a hydraulic test. 'He were a Scotchman,' Fred recalled, 'and he knew what he were about; he were a proper boiler inspector. When we did the test it were bloody leaking everywhere. There were blobs of water showing and really they shouldn't have been doing that at 250 psi, which was the pressure we had it at. But the Scotchman said it would be all right. "Once you light a fire in it," he said, "it won't leak." ' The rusting old roller was nothing like the shiny showman's engine that had inspired Fred when he was a lad, but at least it was a steam engine that he could call his own, so Fred bought it. Talking about it years later he said he thought he had been ripped off. He'd paid £175 for it and he found out later that you could buy a steamroller at that time for about £60 and just beat the scrap man. But back then he wasn't worried because he had become the proud owner of his first engine.

Gerald Adamson's place was only about six or seven miles away so they towed Fred's new possession back there. A fortnight later, after various minor repairs, Fred went for a run on his roller down the road. 'It were incredible,' he recalled. 'Everything was falling off it. But somehow we managed to drive it the nine miles or so back here to Bolton. The roller was a wreck. The back wheels leaned in on each other and if you went over a manhole cover the road wheel banged on the edge of the flywheel. All the gearing had been built up with electric welding and it had worn away again. The boiler and the water tank leaked and the engine made the most unbelievable noise that you could ever imagine. In fact that is half the reason why I'm deaf.'

In those days Fred was still living in his mother and father's terraced house and his brother Graham remembers when he first brought the engine home. 'He parked it in the street in front of Mum's house, and believe me you'd never seen anything like it. It was all scrap iron; you wouldn't give five pounds for it. And as for the neighbours, well, they'd not much choice, had they? Can you imagine one of them engines in front of your door, in a row of terraced houses? Then the guy who had a second-hand car showroom across from where we lived told Fred he could park it up on his car lot.' For Fred it was better than having it parked outside the house. 'The only drawback to that,' he said, 'were the fact that you had to sheet it up and uncover it and sheet it up again any time you did any work on it, which was a hell of a job. Anyway, the good thing was nobody damaged it while it were there and I persevered with it and started to get on quite well.'

But the situation was far from ideal. By this time Fred had married Alison, and with a new wife and a new steamroller to look after he couldn't stay at his mother's much longer. He had to get a place of his own. When Fred and Alison were first shown the gatehouse, Fred said, 'We'll take it on condition that I can bring my steamroller up here.'

'Oh Fred . . .' Alison began, worried about upsetting the neighbourhood from the start. But Fred was firm, even though it meant having to cut a new approach road round the hillside.

Before deciding to take the house he'd worked out the lie of the land. 'I thought,' he said, 'I could dig a bloody road down the side of the hill. So me and some lads who lived next door to where Alison used to live dug the road down the hill. Irish guys they were; one of them is a bloody solicitor now. Anyway, we dug the road down and then I knew a bent weighbridge operator at a quarry who gave us about four wagon loads of ballast and we put that down. When the road was done we came from my mother's with the roller one Sunday and backed it down the drive. The trouble was I made a fatal mistake and backed down on to the grass, and both back wheels just went round and made a huge bloody hole that it got stuck in. The grand plan had been just to get it here and then bugger off back out again to the pub for a pint. But we were stuck in the garden all afternoon with railway sleepers and a jack. We were digging all afternoon, just to get it back on the bloody road again.'

Once Fred got to work on it the roller created all sorts of problems. There was so much wrong with it, it was difficult to know where to begin. Fred would spend all day and night trying to work out how to solve problems, like how to cut, roll and drill inch-thick boilerplate, and how to do all the riveting that was needed. To give some idea of the task that Fred had set himself, it meant having to drill 232 holes in the firebox alone, with an ordinary ratchet drill. He'd turn that handle by the hour, with bottles of Guinness lined up beside him to help things along. At least, Alison used to say, it kept him out of the pub. She was right; once Fred got his teeth into a job he wouldn't leave it until he'd got it done. He'd often be tempted to go for a pint, but he knew he'd meet some of the lads down there and he'd finish up having a few pints with them and all the while he would be thinking of those holes waiting to be drilled.

Neil Carney worked with Fred for several years as his assistant. He first met him about 1968, just after he'd acquired the roller. 'A friend of mine said I must come down and see Fred's yard. The roller was completely stripped down underneath a tarpaulin that was slung between two trees. There were none of his buildings then that he had in later years. I was interested in what he was doing because my father was in engineering and I'd been in engineering in one form or another all my life. So we hit it off. He was no boilermaker at that time but he finished up as quite a good boilermaker. That was by virtue of getting to know one or two retired boilermakers and asking them to come round and give him advice. He was always prepared to do anything he could himself, provided someone just showed him what to do. I think he acquired a lot of boilermaker's tools, caulking chisels and suchlike, from these retired chaps. And they showed him how to use them and you didn't need to show Fred very often how to use a tool. If the reason for using that tool was to get a job done that needed to be done he'd very soon pick up the ability to use it.'

Fred never forgot what a terrible state the boiler was in during those early days and how much there was for him to learn. 'It had a great weld under the front like the wart on an ash tree. When you got inside the firebox there were great globs of welding from one stay bolt to another and it used to frighten me so I just pretended it wasn't there and carried on with the beautification of it. I put new lagging round the boiler, a new funnel and all nice painting and lots of lines on it. But it wasn't right. I recall one night when I'd just arrived back home with it, it were something like one o'clock in the morning and I heard all these noises from the engine. I heard this swish noise and I thought bloody hell, what's this? There was plenty of water in the boiler. I bent down and I looked in and sure enough one of these big lumps that were a weld on the side of the firebox was the reason for the swishing noise. The water was running out from it down the plate and as it did it was turning into steam because of the heat. In the morning I emptied the boiler, sent for the welder and got it welded up. Then I filled it up again to see how it was. I was pretty sure it wouldn't explode because there were too many stays to hold it all together and a leak there wasn't serious. They used to drive main-line locomotives with water pissing out of tubes and they would still go without any problem.'

For Fred, working on the engine was like serving a strange sort of apprenticeship. Because engines like this had been built in the Victorian Age, all the work on it had to be done in the way the Victorian engineers did it, which was the hard way. They were very single-minded when it came to getting a job done and it made Fred the same.

He was to need all that single-mindedness when it came to the next job. As you work your way back on a steamroller or traction engine you come to the most difficult bit, which is the gearing. All the gearing on the roller was very worn. Luckily, in the course of his steeplejacking Fred had gained access to a lot of engineering works. So he was able to get a full set of blanks for the gearing from Hick Hargreaves in Bolton for nothing and it only cost him £200 to have all the teeth put on.

What a difference that made for Fred. 'It went from driving a mad machine to one where you could go along the road and actually talk to whoever was with you on the footplate – which was incredible. I remember with all the old gearing the lids on the oiler used to dance about as you went along the road. But when we put these new gears in they just stayed put and they didn't rattle. I also made a brand new tender. I'd made the engine look magnificent and it was better than when it was new but it still had this big wart under the boiler at the front and the firebox was still full of horrible weld and ready for blowing up.

'We did a whole summer season with it, going here, there and everywhere. I knew that I had to face up to the terrible surgery that the engine needed. I won the cup that year for the best steamroller in the Lancashire traction engine club and it just happened that my boiler inspector was there at the show. It was a lovely summer and he had his shorts on and I remember after I'd received the cup and I was feeling quite proud of myself he appeared on the scene. I said to him, "I'm going to take it home now and I'm going to pull it to bits and make a new firebox and a new front tube plate and a new boiler barrel because the original ones are knackered." But the inspector panicked and said, "No, don't do anything until I've seen the superintendent of the insurance company." It all got sorted out in the end – but the first thing needed was a new boiler barrel.

'With a boiler barrel everything has got to be done by

people who are used to making pressure vessels. I went to a works and they rolled me this piece of plate round and welded the seam. To check it they have it all X-rayed and I was present when they did it in their metal laboratory. I asked the man if the weld was all right and he said, "Yes, there's nowt wrong with that weld. There's no cracks and no muck in it. You might as well take it home and get on with it." '

So Fred took the boiler barrel back to Bolton and drilled about ninety holes in it with a ratchet drill, which he always referred to as the creation of the devil because it was such hard work compared to a big power-driven drill. While he was working on it the phone rang and it was the boiler inspector. 'I should hang fire a bit,' he said. 'The elongation on the weld metal is a few decimal points short of what it should be and we cannot accept this.' The people who'd done the weld for Fred were adamant that there was nothing wrong with it. Their standards were high because they did a lot of work for the North Sea oil people and they got their radiographer in and their examiner and he also said there was nothing wrong with it. The Scottish boiler insurance company and the welding company argued between themselves and it went on and on and Fred was getting nowhere. In the end the people who did the welding said that if the insurance company was not satisfied they would grind the entire weld out and reweld it for nothing. But they wanted another £72 to examine it, which in those days was too much money for Fred, so he got in touch with the insurance company and said, 'What happens if I rivet it like they used to in the old days?' and that was different. They told Fred that he could rivet it and if it stood the pressure of the hydraulic test when it was steamed and the holes were drilled at the right centres, they would accept it. Looking back on it, Fred could understand the insurance company's position. 'It must have been a bit difficult for an insurance company back then when a young man came along and said to them, "I'm going to make a new boiler in a shed in the back yard!" After all, I was a steeplejack, which is miles away from being a boilermaker. There's lots of people doing it now all over England and they all stick to the rules. You've got to stick to the rules of boilermaking. But it was different back then in those early days of steam restoration.'

When it came to boiler-making Fred was lucky, because he lived in close proximity to the railway town of Horwich. Between 1870 and 1965 the town was strewn with miles of track and very long works for the busy production line. Right up to 1957, nearly 2,000 main-line steam engines were built there and a further 50,000 repaired. Although the works had closed by the time Fred was rebuilding his engine, there were a lot of men still alive who'd made locomotive boilers for a living. Some of them were so old they couldn't really do the work themselves; they'd just tell Fred how to do it. But he did manage to find two good teachers who came along to give him some hands-on advice. 'I know one thing,' Fred always used to say. 'These men gave me more knowledge on a Saturday morning than you could have learned from a thousand books. You can read a million books about things like riveting but there is a big difference between reading and having to do it yourself. All these skills were shown to me by these two lovely old men who smoked Capstan full strength and coughed profusely all the time while they were doing it.'

With their help and advice, Fred made the new boiler. It was almost finished and he'd got to the hydraulic test stage. This involved filling the boiler with water and using a hydraulic pump to pressurize it to one and a half times its normal working pressure under steam. As he was testing it along came the boiler inspector.

'It's got a blob – a leaking rivet – not pouring out, just wet,' said the inspector.

'What can we do about it?' Fred asked.

'Well, you could put a run of weld round it.'

Fred wasn't very happy about this. He'd made the entire boiler using the old-fashioned skills and now this guy was telling him to put a weld round it!

'I don't want to do that,' Fred told him. 'I never wanted it to be welded.'

'There's only one other way that you'll stop it and that's to fill the boiler up, empty it and leave it standing to let it rust. That's the way they did it in the old days.'

But back then engine builders had time built into their schedules for this. Fred didn't. He wanted to have it on the road within three weeks. So he had a problem.

But Fred was nothing if not inventive and, after hours

and hours of lying in his bed thinking and thinking about it until he thought his brains were going to explode, he cracked it. He came up with a simple solution that involved putting an iron bar across the rivet where he'd got the leak. About ten minutes after he'd fixed it up, there were no leaks. But then he had another problem and this one was much more serious. It was financial. The Scottish boiler insurance company sent him a bill for £175 for the supervising they had done so far. Fred didn't understand this. 'They weren't really doing anything, just creating obstacles all the time.' So he paid the bill and sacked them.

At that time Fred was working at a mill where they had just had some work done on the boiler. 'Did you do a hydraulic test?' he asked the engineer.

'No, no . . .'

'What about the boiler inspector?'

'Well, the boiler leaked but the inspector said, "Oh, it'll be all right. It'll seize up as time goes by and the drips will eventually stop."'

Fred decided that he could do with a boiler inspector like that, so he got his name and phone number from the engineer at the mill. Mr Farrington was his name and he lived near Horwich Loco Works. Fred gave him a ring and instead of saying, 'I've got a locomotive boiler in my shed. Can you come and examine it?' he told him the entire tale about how he'd made the boiler in the back yard.

Mr Farrington said, 'I'll see you next Wednesday at

eleven o'clock but I'll have to talk to the superintendent first.' Next Wednesday arrived and at a quarter to eleven the phone rang and the man at the other end said, 'It's Farrington here. I've had a word with the superintendent and he said, "Stay away from it. Don't go near it!"' So that was that. Fred had no boiler inspector.

At the pub that night with the railwaymen and his boiler-making friends, he discussed the merits of individual boiler inspectors. One name kept cropping up – Kojak, a big man with a bald head who wouldn't let anything pass. If the slightest thing was wrong, he wouldn't have it. By this point Fred was desperate, but he didn't like the sound of Kojak. He had about a fortnight left before he wanted to have the roller out on the road. He knew the thing was all right. There were no leaks and it had passed the hydraulic test. So he rang up the man who insured his Land-Rover and the man said, 'We don't really do industrial stuff. We only do motor cars and houses. But leave it with me and I'll get back to you later on today.' Later on that same day the phone rang and he said, 'Next Thursday somebody will be coming to see you about your boiler.' On the appointed day a big car pulled up outside Fred's house. Out stepped a man with an executive leather briefcase with brass corners and an expensive-looking suit. Fred had got steam up and had the engine blowing off with both safety valves roaring into the sky, creating the type of noise that used to frighten small children on railway stations. The man with the briefcase walked slowly round the engine about three times and said, 'Have you made that?'

'Yes, I've done nearly everything apart from rolling the plates,' Fred replied.

'Well, it's a credit to you,' he said, 'but I'm not a boiler inspector so—', and his next words filled Fred with fear— 'Mr John Gourley will be coming to see you next Tuesday.' This was Kojak – the man dreaded by all the boilermakers in town.

Fred never forgot John Gourley's first boiler inspection. 'That man was the most thorough man that I've ever come across.' After Mr Gourley had finished with the hydraulic tests, the next part of the inspection was to empty the boiler. He looked inside it with a mirror on a stick and a little torch to spot any grooving or any wasted stays and to examine the state of the smoke tubes through the actual water space in the boiler. Then after he'd done all that he said to Fred, 'I'm really sorry, I can't come back and see the engine under steam for at least three weeks because I've got a busy time doing all the mill boilers.'

It was another blow for Fred, but he knew Mr Gourley lived in Bolton and, quick thinking as ever, he said, 'By the time you've had your tea I can have this full of water and steaming. Could you not come back after you've had your tea?' He agreed to this and asked Fred if he could bring his son back with him. After tea they both arrived and by this time Fred had the engine blowing off with no leaks and all the rivets tight and good. The first thing Fred said to them was, 'If you've got ten minutes I'll take you for a ride on it – over the cobblestones, so we can prove it's not held together with chewing gum and goo.' So they set off on the steamroller over the cobblestones down past Burnden Park football ground and down through Rakes Clough on an unmade road full of pot-holes and great ruts. Then they stopped at the Lever Bridge pub, had a couple of pints and arrived back at Fred's front door. Mr Gourley said, 'Your ticket will be in the post in the morning.' And that was it. From that day on he became a firm friend. 'He used to bring round trainee boiler inspectors,' said Fred, 'and when he retired from British Engine in Manchester he invited me to his retirement party. I learned a lot from him – he was a good man.'

John Gourley was just as impressed by Fred and he remembers that first meeting equally clearly. 'I was an engineer surveyor employed by Apollo Insurance Company. When I went to carry out the first inspection at his house I recognized almost straight away that there was something unusual about this chap. I had examined many restored boilers and engines and bits and pieces like that, and when I came to meet Fred I realized he was an extremely thorough person. The preparation of the boiler was beautiful and the repairs he'd carried out were excellent.

'I thought he had quite an amazing garden and work-shop. I was most impressed with all the machine tools he'd assembled there, and the structure to drive them. This was made out of telegraph poles and would bring about many problems with shaft alignment. It was quite a scanty kind of structure really, but he'd made a very good job of it and it worked perfectly. I liked the fact that he could do all the work himself without much assistance.

'Over the years I had to inspect all of Fred's engines and boilers. And if I could describe the whole set-up in a word, I couldn't fault it. He'd taken meticulous care with every bit of the restoration of his engine. I've looked at many restored engines and boilers, and I can honestly say I think this was the best I've ever seen. He wasn't even trained as an engineer! But he was trained as a joiner and that calls for accurate workmanship, to be able to work to drawings. He picked up the detailed engineering skills by watching and listening to other people. He was a very good listener and a good watcher, and he was able to hone his skills on these engines.'

Like Fred, Michael Webber had a passion for steam from the time he was a small boy growing up in Bolton in the 1970s. He now has his own steam engineering business but his apprenticeship was with Fred. 'My father was a school teacher, and he used to take us to his school every summer for their summer fête. Fred used to take his engine along so that's when I first met him. At that time Fred was only interested in going to local events and every weekend he would take the engine somewhere. It might have been just a few hundred yards down the road, but he always liked to take it out for a run.'

Michael was fascinated by the engine and, as there was

always a lot of work to be done on it, Fred was happy for him to give a hand. 'My association with Fred was very strange really,' Michael recalls now. 'Obviously a young lad has many interests, but Fred just had something about him that was very appealing. It was exciting, being on the engines and being dirty and having people looking at you. Gradually I started going to the house more and more and I got to know Alison very well. Alison approved of me and she was supportive of me helping Fred with the roller. Every time we went out on the engine it would be quite hard work – steaming up, on the road, not eating properly during the day – so when we got back to the house Alison always had a big plate of egg and chips ready for us.'

Fred liked having Michael around to help him because he would have liked a lad of his own to play with him on his engine. By this time his family had grown, but he'd not yet had a son. 'Alison and I were blessed with three daughters – Jayne, Lorna and Caroline,' Fred would say, 'so I had a gang of four females around me who expected to eat, to be clothed and even to go on holidays.' But holidays for Alison and the girls tended to be weekends away at traction engine rallies. Michael Webber would often go with them.

By this time Fred had taught Michael how to drive his engine. 'He was quite confident I could work it on my own if he wasn't there,' Michael recalls. 'This was good news for Alison because it meant that when we'd been out on it I could take her and the girls home on the engine and leave Fred in the pub with his mates.

'At that time there weren't many rallies around Bolton. The nearest of the big summer rallies was at Burtonwood near Warrington, which was probably twenty-five miles away. Because Fred's machine was quite slow it used to take us a whole day to get there. Back then I didn't drink and Alison didn't really drink either. In those days pubs didn't sell food like they do now. All you could get was a bag of crisps or peanuts, so I was always starving hungry. Fred used to start steaming up at about six o'clock in the morning. We'd then be able to set off from the house around nine o'clock and we'd get down to the Lever Bridge pub by opening time. There Fred would get his first three pints and then we'd trundle off for another hour or so to the next pub. When people asked us how fast the engine went I'd say it used to go two pubs an hour. And that was what it was. We used to stop and start all day and I'd count how many pints Fred used to drink and, believe it or not, he could get through sixteen during the course of that journey from Bolton to Burtonwood.'

For Fred there was a good reason for all this drinking. 'Because of their speed, or lack of it, steamrollers take a long time getting anywhere. The funny thing is most of us steamroller and traction engine men find driving these monsters a lot easier when we've got a few pints inside us than when we're stone cold sober. Drink sort of eases the pain. It feels less bumpy under the iron wheels, and the racket that's always going on all around you subsides a bit.'

Fred remembered one instance when he was driving the engine back from a rally all through the night. Along with his steersman he'd been in the pub till closing time and then they'd got a bag of carry-outs to ease the pain as they trundled along through the darkness at four miles an hour. His steersman became so drunk that he fell down at the back of the engine. He was a giant of a fellow, well over six feet tall, and he wore size fifteen boots, which stuck out at the end of the tarpaulin that Fred hastily threw over him. Fred was chuffing along quite happily doing the steering himself, when all of a sudden a couple of police cars appeared and skidded to a halt just in front of the engine. Half a dozen policemen got out and came towards it. They were all looking very serious, with the sergeant in charge the most serious of them all. It didn't look good. Fred had a think, and before that very serious-looking sergeant could see the condition he was in and all the bottles rolling around the floor, not to speak of the unconscious figure lying in the back, he had a flash of inspiration and said to the sergeant, 'Do you like these things?' Fred never forgot his reaction. 'You should have seen his eyes light up. He was like a little lad all over again, a boy with a new toy.' So Fred said to him, 'Would you like to have a go?'

His eyes went even bigger and he asked nervously. 'Can I drive it?'

'You can't drive it,' Fred said sternly, 'but I can let you steer it, under my supervision.'

'Of course, of course,' the sergeant said humbly, then

helped Fred to shift his steersman, who'd begun snoring. By this time all the other policemen were smiling. It made a nice change from real crime, to come across a minor Victorian felon like Fred. The sergeant and Fred drove about four miles, which took them an hour. They steamed up a long hill into town, with the rest of the squad escorting them. By then it was dawn, and the sun was rising. It shone on the sergeant's face, and Fred said he'd never seen anyone looking so happy.

Unfortunately steam engines didn't give pleasure to everybody, especially irate motorists in a hurry to get somewhere. But Fred was never in a hurry to get anywhere and he thought motorists were as big a problem for him as he clearly was for them. 'The main trouble with taking a lovely old steamroller for a run,' he would say, 'is that the roads are full of motor cars doing incredibly stupid things. They don't seem to realize that, though I can stop or start my roller just like their motor cars, it must be done in a careful kind of way, and things all take a bit longer. It's always a bit tricky when I'm approaching traffic lights because they seem to attract motorists whenever there's a steamroller ahead. As they are coming up to the lights the car drivers feel they need to get in front of the steamroller so that they can move off smartly as soon as the lights change, so they nip in front just as I'm approaching the red light. What they don't understand is that I can't see exactly where they are while I'm trying to bring the roller to a stop. If they knew the peril they were putting themselves into, a lot of these drivers would think twice before trying to get between me and the lights.'

The attention steam engines attract was another source of trouble for Fred. 'As they shoot past me in their motor cars, most drivers turn round to have a look. Once, I was going up a long hill and at the top of it there was a bus standing at a bus stop. The driver of the car that overtook me was so interested in my roller that he drove right the way up the hill looking round all the time, and he finished up running into the back of the bus.'

Fred would get a queue of cars behind him and there'd always be one impatient fellow hooting away somewhere along the line. That would make him a marked man, as far as Fred was concerned. He knew that the driver was going to try to overtake the whole line on the next straight stretch and he also knew that the man would toot his horn as he was accelerating past. Fred would keep an eye on him as he pulled out and came racing along on the wrong side of the road, then, just as the car was going past the roller, Fred would blow his steam whistle. At 200 lb per square inch, that whistle would be quite a shock to anybody close to the engine.

For some people, driving a steamroller or tractor is a particularly pleasurable experience because its gentle pace gives plenty of time to see the countryside. But to Fred, driving through the countryside was rather boring, for the beauties of the British landscape meant nothing to him. 'I was born and bred among streets, factories and chimneys,' he would say, 'all of which are interesting and different, while one field looks just like another to me and so do trees and hills.'

For Fred, a steam engine was special. 'A steam engine,' said Fred, 'gives you a sort of immortality. When you resurrect one from a pile of rust, you put your own stamp on it. The thing is, when I'm dead and gone, it will still be here. You've got to have a very good wife, though, if you have a steam engine – one that understands. When I got married, things were a bit rough as far as steam engines were concerned, and divorce procedures were imminent on quite a lot of occasions. But as the years rolled by Alison got to like the thing.' Jayne, Lorna and Caroline grew up with their dad's engines and had to get to like them as well, because they chuffed off with him to every steam rally within driving range. There was never any other kind of holiday for the young Dibnahs.

For family use Fred needed a living van with beds and cupboards and cooking and washing facilities like the old steam-roller and traction engine drivers used when they were working away from home. Although Fred was still very short of cash, he managed to acquire a van in his usual manner. As he went about doing his chimneys he kept his eyes open and he spotted one in the middle of a field near Burnley. He went to the farmhouse and found out that the farmer was planning to use it for firewood, so Fred offered him £35 and came away with a bargain.

The van was in poor shape but it was mainly joinery work that was needed and Fred was well qualified to do that. But it took him a lot of time and a lot of wood and for a while it diverted him from work on the roller. When it was finished he started to tow the van behind the roller with his young family on board. It was a support vehicle cum mobile home with enough beds for everyone, and a huge coal compartment that you could get to from the outside. The girls would travel in the van and, as they got a bit older, they progressed to sitting on the front footboard. Perched beside them was their dog, Nickie, a little white mongrel. The dog liked the steamroller and, come rain or shine, would always perch on the footboard. As they rolled along he changed colour, from white to dirty grey.

Doing a long journey on a steamroller or traction engine has problems all of its own that you'd never have to think of in a car. For a start, such vehicles need plenty of coal and water, and keeping the water tank full could be difficult. Right from the beginning Fred dealt with this by taking water from hydrants along the way. Fred's roller would do seven to eight miles on eighty gallons of water, so he had to take on a supply whenever he could. If the level were to go too low the boiler would explode, endangering his life and that of any passer-by.

Another thing you have to learn to live with when you're driving a steam-powered vehicle on the road is the noise. If an engine has worn gearing it's impossible for the driver to speak to anyone more than a foot away, except in sign language, and people can hear you coming for miles. Fred's engine wasn't too bad, but years of driving it all over the place did leave him deaf in one ear.

Then there was the oil problem. The drops of oil flying around got in your eyes, making them sore. But on the plus side, it was a good way to tell whether the engine was running properly or not. If the drops of oil being sprayed into your face turned from nice and warm to red hot you could be in for some real trouble.

Too much sunshine was another hazard. After a long run on a sunny day, everything became too hot to handle. Even the footplate turned into a frying pan and your feet would roast inside your boots. That was why the old-time steam engine drivers always preferred to travel in the rain. They'd be under a decent-sized roof that would protect them while the rain on the rest of the engine would act as a water-cooling system. Unfortunately, rain can cause another problem for steamrollers. Because they don't have rubber tyres, the wheels and the rollers can slip and skid on wet hilly roads. 'Going downhill can be bad news,' said Fred, 'especially if there are metal manhole covers. Once a steamroller starts skidding it becomes a monster.'

Fred had a good policy of making sure he used the engine and had fun with it every summer, and then he'd do any work that needed doing in the winter. So throughout the winter months it was always in bits. 'That was good for me,' recalls Michael Webber, 'because I wanted to learn about engines; how all these big gear wheels and pistons and things worked. In the winter time there were always lots of bits for me to look at, and in the summer it was all get it going and use it, use it, use it. We'd do miles and miles on the road. Fred always wanted to use his engine, so it was not stuck in the shed just to be looked at. It had to work and it had to be used.'

As he went to the rallies Fred met fellow steam enthusiasts who were to become lifelong friends. One of them was Dick Ransome. 'I first met Fred at Burtonwood in either 1969 or 1970. It was quite an exciting meeting because it was in the middle of a thunderstorm and Fred had his roller there so he was soaked to the skin. He'd only just finished the engine and he hadn't had time to lag it properly so the boiler was just lagged with sacking, which was all steaming with the rain. Of course we finished up in traditional form in the beer tent and that was the start of a friendship that was to last for thirty-five years.'

Fred spent thousands of hours playing with his engine, to the detriment, as he confessed, of other aspects of his life. 'I do spend more money on steam engines than on the wife, kids and holidays all put together. A lot of people have said I spend too much time playing with steam engines. "You should come and get our job done," they used to say, which was very true. I have neglected my business and everything, really, for the sake of this ten ton of iron.' So it was bad news for Fred's family and for his customers when, in the spring of 1980, he was spotted building an extension to his back-yard engine shed. Their worst fears were realized when it came to light that he had his eye on the rusted wreck of another steam engine.

5

Changing the Landscape

By the late 1970s Fred felt he had reached the pinnacle of his steeplejacking career when he got the job of repairing the clock tower on Bolton's magnificent Victorian town hall. Right up near the top in the balustrade of the lantern there were sixteen stone pillars. After years of exposure to the smoke from the town's mill chimneys some of them were in pretty bad shape. Fred was doing some work on the town hall at the time so he drew the attention of the Clerk of Works to the state of these pillars. He got in touch with a firm in Stockport who said they would make some replacements in fibreglass. But Fred cringed at that. 'Fibreglass on the town hall? Let me have a go at making some in stone,' he said.

From the local stone merchants Fred purchased three big lumps, each about three feet long and six inches square, to replace the worst of the pillars. He knocked up a lathe in his back yard from an ancient angle grinder and

150

some abrasive discs and, with the motor from an old Hoover to drive it, he carved the pillars into shape. When they were finished Fred was proud of his work; his new pillars looked magnificent. He put one of them in a sack, walked into the town hall, set it on the Clerk of Works's desk and said, 'How about this? Don't you think this is better than fibreglass?' So Fred got the job of restoring the pillars and it opened up what was to become a useful sideline for him, the stone-cutting business. After he'd replaced the stone pillars he talked himself into the job of gilding the ball right up on the top and pointing the whole lantern – but not before an interesting development that was set to change Fred's life.

Swinging past the clock face in his bosun's chair high above the shoppers in the town hall square, Fred was enjoying himself. 'I were mending the golden ball on top of the flagpole, putting the gold leaf on. It's a lovely business sticking gold leaf on, especially on a summer's day when there's no wind. I was up there getting on with the job when along came the town hall's public relations man. "I've got a bit of a surprise for you," he said. "Next Thursday, *Look North West* are coming to do an interview with you."'

Look North West was the BBC's local news programme and the idea of being on television was quite frightening to Fred. ' "You'd better smarten yourself up and get a clean shirt and all that," the PR man said. Anyway,' said Fred, 'the next week the big grey van arrived as arranged with what seemed like two hundred people from the BBC. Alistair Macdonald was going to do the interview and he had a reputation for being a bit of a mountaineer and a rock climber, so I thought this would be right up his street. I reckoned it would be a very good interview because when he got up on the roof he wouldn't be holding on to everything, scared stiff. The television team got the lift for two floors, then it was an iron ladder or two and through the clock before entering the lantern and eventually out on to my woodwork. When Alistair Macdonald arrived at the top he'd got his clipboard in one hand and a mike in the other. I said, "Are you coming on to the woodwork to do the interview?" I'll never forget his reply. "Bugger off!" he said. "No way! I'm staying in here." He wouldn't come outside on to my scaffolding, so that were it. He did the interview from inside, shoving the microphone through the holes. That night I was on the news. They always put the idiots on last and that's where I was. I watched it and they didn't make me look too much of an idiot. And that's how it all started.'

A few weeks later Alistair Macdonald rang to ask Fred if he could come to see him with a television producer. 'Why not?' said Fred. Being on television wasn't something he'd given any thought to up to this time, but if somebody wanted to make a programme about his steeplejacking it wouldn't do any harm to business. Fred never forgot the telly folk when they first arrived. 'This guy with Alistair looked like Stewart Granger, about seven feet tall, all dressed in denim, baseball cap and huge cowboy boots. We chatted, or rather he talked and the rest of us listened, but it was mainly media language, which I couldn't make head nor tail of. At the end of our chat I said, "What do you want me to do, give you a ring when we're going to do something dangerous or exciting?" "No," he said, "we'll ring you." And that would be the end of my television career, I thought.

'Normality had almost returned to my life when Alistair Macdonald, who could never be called a quitter, had a hand in things again. He'd been pestering people at the BBC in Manchester. As a result of his efforts I received a call from a very nice lady called Jean Thompson. We arranged a time to meet and she arrived with another Stewart Granger character, only this time a smaller version, and much older, but the same kind of boots, a long peaked cap and a nicely tailored denim outfit. It must have been some sort of uniform that the telly types wore back then. I attempted to show him, with the help of a pencil and a sheet of paper, how steeplejacks get a ladder fixed up a chimney. We had about one and a half imaginary ladders up this imaginary chimney, when he waved his arms. "Stop, stop! We'll have the viewers turning off in droves. That's much too boring." Then he had a quick chat with his lady colleague and he was gone. Blown it again, I thought. The idea of me being on telly seemed quite unlikely at that time, but if there was going to be a programme about my job, the more I pondered about it, the more I was sure that the first part of any programme would need to contain the method of putting the ladders in place.'

The smaller Stewart Granger character was BBC producer Don Haworth and that first reaction of Fred's was to be the same throughout all the filming he did, not just with Don but with me later. In total Fred appeared in more than sixty documentaries for the BBC and he always felt that the most important thing was the level of technical explanation. The only complaint I ever had from Fred in seven years of working with him on forty films was 'All the best stuff ends up on the cutting-room floor.' But, just as Don Haworth had explained to him about the laddering, I had to keep letting him know that the majority of our three million-plus audience didn't want long and detailed explanations of the workings of the mechanical lubricator or the relative advantages of twin tandem compound engines and triple expansion engines. I spent many hours, usually locked in various pubs after closing time, listening to Fred's explanation of how he laddered a chimney and, fascinating as it was, it wasn't the sort of thing to pull in the viewers.

With the first film, Don Haworth wasn't going to be put off. He knew that climbing up and down chimneys and driving round on a steamroller wouldn't fill a fifty-minute documentary without a very colourful and engaging character at the heart of it. While Fred wasn't exactly Hollywood hero material, Don Haworth thought his flat cap, flat vowels and eagerness to tackle hair-raising challenges would fascinate millions of television viewers. Above all, a fifty-minute prime-time documentary needed somebody who could talk and at their first meeting Fred had held forth for two hours, so Don had found his man. Fred got a phone call. It was Jean Thompson, research assistant and later associate producer, from the BBC. 'Can we come over on Wednesday?' They did, and according to Fred they had another argument about the ladders.

'But all of a sudden,' said Fred, 'Stewart Granger Mark Two showed signs of capitulation.'

' "Where are you working?" he asked.

' "On a big chimney at Shaw, near Oldham," I told him.

' "Jean, we'll call there tomorrow and have a look for ourselves," he said, and off they went again.'

Fred recalled that he'd just got to the top of the 245-foot chimney the next morning when the BBC pair turned up in their Mini. It was 8.30 a.m. 'That was it,' said Fred. 'Cowboy Boots instantly got the bug and they started filming.' For Fred it was all a bit of a strange experience. 'They used to film for two days and then they'd disappear for three weeks and I'm thinking it's a funny way to make a film, this. Then they would come back and do another day or two and disappear again and all this went on for eighteen months for one programme that lasted fifty minutes.'

The film that Don Haworth made in 1979 showed Fred at work, erecting scaffolding around chimneys two hundred feet up, and Fred's antics up those chimneys made compelling viewing. But the real drama in this and all of the early films came from the demolitions. The first one was filmed one Sunday in 1979 in Rochdale and it was this, more than anything else, that set Fred on the road to fame. Hovering nervously near the base of the chimney, he would signal the end with a couple of toots on an old motor horn he carried with him. As the chimneys crumpled in almost tidy heaps – to the accompaniment of his cries of 'Did you like that?' – he struck a warm chord with viewers.

Don Haworth's documentary, *Fred Dibnah, Steeplejack*, is widely acknowledged as a classic. It won a BAFTA award for Best Documentary and in a recent Channel 4 programme on the most influential documentaries ever made it came in the top fifty. It brought Fred instant fame, turned him into a minor celebrity and led to a host of other programmes over the next twenty-five years that made him one of the most famous names on British television. One thing that wasn't in the film was Fred's explanation of laddering a chimney, though he'd tried to hold out for it. If the sort of lengthy explanations that Fred wanted to give had been in, it's more than likely that *Fred Dibnah, Steeplejack* would have been the first and last we'd have seen of Fred on our screens. What is more, it was only because of the television climate of the late 1970s, when we still had a strong tradition of one-off documentaries that didn't fit into any particular slot or 'brand', that a film like this could be commissioned and shown on a major channel. If Don Haworth were to go along today with an idea for a film about a rather eccentric character who earned his living as a steeplejack and spent all his spare time restoring old steam engines, it's very unlikely that it would be commissioned in today's format-obsessed world of television. And how much the poorer we would have been without twenty-five years of Fred on our screens, keeping us entertained and informed of the things he was passionate about as he held forth on modern society.

The late Barry Cockcroft, who made *Too Long a Winter*, the film that first brought the Yorkshire Dales hill farmer Hannah Hauxwell to our screens, said in the late 1990s, 'Television has lost the art of making ordinary people extraordinary. The medium is much more competitive now and programme makers are able to spend less time on individual documentaries. You need patience to make the people you are filming oblivious to the presence of the camera.' Since then it has got even more difficult to make this type of documentary, as programme makers now have to tailor their ideas to the slots that commissioning editors and channel controllers have to fill. The constructed reality of shows like *Big Brother* or *Wife Swap* is a far cry from the reality of *Fred Dibnah, Steeplejack* or *Too Long a Winter*, as is the ordinariness or naturalness of their participants compared to Fred or Hannah.

By the time Don Haworth came into his life, Fred had a new assistant who was set to become almost as famous as Fred through his appearances in the films. Donald Paiton was Fred's longest-serving assistant. A one-time builder with a grammar school education, he had come to work with Fred in the late 1970s, when he was in his sixties. Donald was a humorous man with a very sharp intelligence and he became a great foil for Fred, just as Alf Molyneux was to become in some of the later films.

Fred was always proud of that first film. He saw it as his film and used to tell everyone that he'd won 'an academy award' for it. After its first showings he began to find that strangers recognized him. He'd always liked to talk and amuse people and his new-found fame started to give him a ready audience for his stories and his views on the world. And to add to his overall satisfaction with life, he'd just won the cup at the Burtonwood Vintage Show with his roller, so that brought steam into the plot.

However, even for an enthusiast like Fred, steamrollers did have shortcomings as a mode of transport. What he'd always wanted was a traction engine with proper springs and rubber tyres and proper wheels on the front instead of a two-ton roller. A good traction engine would streak along the road at a very respectable fifteen miles an hour rather than the snail's pace of the roller. So by the time the first film was made Fred had got a new passion in his life.

He'd acquired a 1912 Aveling & Porter Convertible traction engine. The compound had originally been built for Somerset Council and configured as a tractor, but by 1914 it had been commandeered by the War Department and sent off to France configured as a road roller. It came back as war surplus and Devon County Council used it for the rest of its working life as a roller. It was parked in a lay-by in the mid-1960s when someone stole all the brass fittings from it. It was then advertised for sale and tenders by sealed bids were invited. The successful bid came from Peter Froud, a friend of Fred's, who had a number of steam vehicles and boats. He got it for only £200. At the time Fred said he was green with envy, having just paid £175 for his roller.

Fred had known Peter Froud for a number of years, having first met him at a steam rally when their engines had collided after they'd both had a few drinks. After a bit of haggling, Fred paid £2,300 for the traction engine and moved it to his workshop. Straight away Alison had misgivings about it. 'It needs a lot done on it,' she said to Fred in one of Don Haworth's films. 'How are we going to find the money?' Fred's answer was quite simple. 'We've managed before,' he said. 'No more holidays.' As the nearest thing to a holiday they'd ever had was a weekend with the roller at a steam rally, Alison got the message. But if there is a point where a hobby becomes an obsession, Fred failed to see it. Nothing was allowed to get in the way of his passion. Colin Shaw, his friend from primary school days, stayed in contact with Fred throughout his life and knew him well. 'I think he was just selfish, to be honest with you. If it suited Fred that was fine, but if it didn't suit Fred, Fred wouldn't do it. This didn't make you dislike him at all, it was just the way he was.'

When Fred got the engine back home he thought it looked very good, with nice lagging on the boiler, belly tanks all lined out and a smart paint job. But there were telltale signs for all to see, including a big L-shaped weld on the side of the boiler right underneath the high pressure valve spindle. The standard of the riveting on the boiler also left a lot to be desired. It quickly became obvious to Fred that this boiler barrel would never pass muster, and so the twenty-seven-year slog began to get the traction engine fully restored.

Fred geared down his steamroller to drive the machinery he'd use to beat out parts for the new tractor. The roller had taken fourteen years to do up. Fred reckoned, with what turned out to be the wildest optimism, that he'd whip through this job in less than four. He worked in the shed sometimes until one o'clock in the morning. Then to make life even more difficult for himself he decided that, in order to get the traction engine done in the way it should be done, he would build a complete steam-driven workshop in his back garden. The thinking behind it was that the more engineering works were being modernized, the harder it was to find anybody to do the sort of jobs that needed to be done on his tractor. With his own workshop, he reasoned, practically all the jobs on his roller and on the tractor could be done on his own premises. But more than anything it was because of his passion for steam.

To house his two engines and all the machinery he was beginning to accumulate he rebuilt and extended his engine shed, using recycled materials. It meant that much of his time was spent hunting round demolition sites for old wood and doors and window frames and scouring scrap yards for all the bits and pieces of machinery that he needed. He then got hold of belts and line shafting from the mills where he was working. Finally he found a small, stationary engine to drive all the machinery from a mill in Chadderton, near Oldham, where he was repairing the chimney.

Alan Crompton watched as Fred put it all together and would often give him a hand. 'In his garden he could produce anything. He'd got all his machines and, of course, he didn't pay much for them – if anything at all – because they were always archaic, belt-driven machines that were not wanted by anyone. But it were ideal. It suited his steam, which drove to every corner of the yard with belts and that. He had drilling machines and then he built a machine for cutting stone with a diamond wheel. It were all round the back of one of his sheds. He could do anything. He were wonderful with metal or stone or wood. Incredible.'

But progress on the new traction engine was slow, because Fred was finding too many distractions. Building his workshop and scouring the scrap yards and demolition sites for his machinery took time away from building the engine. But what was really eating into his time was the fact that he also had a lot of work to do on the house, building a two-storey extension to provide more room for his growing family. 'The thing is, with it being a listed building, everything's got to look just right,' said Fred. 'So you've got to have second-hand bricks. After many a month of looking around at various buildings that were about to have the big hammer, we found at least ten houses in one row. Six of them had never been messed about with, never been pointed or anything, and the bricks were a perfect match for the existing ones. So we went with two Land-Rovers – the wife, me, Donald and my father-in-law and another lad – and we literally took the front out of six of them and brought it back here to build the place. And then the lad who did the drawings got carried away with all this fancy stonework, and stone's expensive stuff. But we had another windfall. We met a vicar who said, "I've got all these gravestones here, and you can have them if you'll shift them." So they're now the window heads, the mullions and the window bottoms. The thing is, I was a bit worried in case the bloody place was going to be haunted when it was finished.'

Fred's brother, Graham, was a bricklayer so he soon got roped in. 'I went and helped him with it. Well, I'm saying helped him, I virtually built it, but I never got paid. I never got a penny. I never got a drink from him for it either. I don't think our kid has ever bought a round!' But everybody who saw the extension admired it. Alan Crompton was a frequent visitor. 'Everything Fred touched had artistry about it. He just couldn't help himself. It was as though it was somebody who were controlling him. Wonderful, the work he did. When he sent the plans in to Bolton Council, they were so impressed that they didn't send anybody down to look at it. They just said, "Fred, carry on with it," because his drawings were immaculate and he knew all the building regulations. Some of his brickwork and his stonework on that extension was magnificent! Then there was the chimney. It was a hexagonal chimney in brick that he got from a house up Chorley New Road. It must have been a real centrepiece on this house and they'd demolished it for some reason. Fred brought it back brick by brick to his house in Radcliffe Road and built it on top.'

But now that he was on television there began to be other calls on Fred's time. The whole celebrity business started to eat into the hours he was able to spend in the shed, particularly as he got curious onlookers coming to see this strange garden they'd seen on the telly. As well as this, fan mail started to arrive, which had to be dealt with. And Fred still had a living to earn. But now there was the added complication of having a lot of his steeple-jacking work filmed, and the views he expressed as he climbed up towards the heavens and swung himself on to his staging two hundred feet from the ground began to grip the nation.

'It's not so pleasant on a Monday morning, when it's cold and the wind's blowing and you look up and you think, oh, good God!' he said in one of the films. 'This year it's been windy nearly every day. I don't think we've had above a month when the weather's been really decent. It's either been blowing a force nine or raining, or snowing or freezing cold. So you suffer. Summer's best, when the sun's shining, and everybody down below is sweating away inside and you are up there with a nice cool breeze blowing – beautiful. This particular job, I've got the contract for seven thousand pounds to knock it down, a brick at a time, right down to the bottom. The reason for the price was that I didn't really want the job. If it had been a repair job, which on average takes a month or five weeks, it's not so bad, but when it comes into months and months, it's a bit different. Furthermore, when there's only one of you, and such a gigantic pile of bricks, you've got to have a stout heart to take it on. There's too many buildings close by to drop the thing, or blow it up as some people do, so I've got to knock it down a brick at a time.

'Really you're dicing with death, with a rotten old top on the chimney. There's been a lot of men died fiddling with them things. I've never fallen off a big chimney; you only fall off one of them once!

'Donald, my labourer, he's a very important man. You've got to have a man with a reasonable amount of grey matter. If you get some young lad, his heart can be in the right place but if he sees a girl in the office he wanders away and when you need him, you look down and he's gone. In the past I've had lots of labourers. Some drank rather a lot and led me astray. Donald, he's a staunch teetotaller so he's a good ambassador for me. He keeps me out of the pub.

'If you worry all the time about the job, you would fast become a bloody nervous wreck. I just don't think about it. The only time when the wind gets up a bit is when you're going to do something that's particularly hazardous. Everyday run-of-the-mill steeplejacking, like climbing up ladders and dancing about round tops of chimneys, is pretty safe. But you've always got this feeling at the back of your mind that if you put one foot wrong you're dead. That's it, you've had it. It's now for a date with the undertaker.

'I know one lad who did fall off. He fell about sixty feet and landed on a load of planks across a valley on top of a building, between two roofs, and of course the plank saved his life. It must have broken his fall, even though it rearranged his bone structure. I found out about this when I rang him one evening to invite him to a chimney-felling operation. His little lass came on and I asked if her dad was in. She said, "No, he fell off a chimney." You shouldn't laugh really. Then Mum came on and said he was in Tameside Hospital. "You can go and see him if you want." I went round and he was all trussed up with wires and strings. He said, "I'm all right until I laugh, and then it feels like someone has hit me in the chest with a sledgehammer." He's all right now, and he's back steeplejacking again.' This mixture of toughness and danger and homespun wisdom kept the audience enthralled and Fred soon built up a devoted following.

To Fred, filming was a strange sort of ritual that he never fully got used to. 'Most of the early films were quite funny really. The crew actually drew lots to see who were going to climb up the chimney to film me there. The man who did most of the climbing was Martin

Lightening and he did very well. It were in the days of sixteen-millimetre film and I got to carry these heavy cassettes of film up and Martin got to the top and did some quite splendid camerawork.'

Whenever he was on a particularly difficult job, there always seemed to be important-looking people superintending things down below. According to Fred, they generally drove Mercedes cars and carried real leather briefcases with brass corners on them. At least one character like this was always around when Donald and Fred were putting up the ladders. When they'd finished laddering a particularly difficult chimney, one of these characters came up to Fred and said, 'I've always been intrigued as to how ladders went up chimneys.'

'How did you think they went up?' Fred asked.

'Well,' he said, 'I imagined they were put end-to-end and connected before being reared up the chimney.'

Donald, who was standing nearby, couldn't help laughing, but Fred managed to keep a straight face. 'It would be a bit difficult,' he pointed out, 'holding your foot on the bottom rung while trying to heave the other hundred and ninety-nine feet upwards.'

'I suppose it would be,' said the chap, which made Fred wonder how he came to own a Mercedes.

Through Don Haworth's films Fred became best known for *felling* chimneys, although that was the job he liked least. His preference was for restoration and he would always point proudly to chimneys and church steeples that he had repaired. But it was the chimney-felling tag that stuck, along with his famous catchphrase, 'Did you like that?' Many people mistakenly believed that Fred 'blew up chimneys' but this was never the way he worked; he had too much respect for the men who had built the chimneys.

'You cannot help admiring the work that went into a single chimney stack,' he would say. 'I've seen them from close up and they're fantastic, great big stones perched two hundred feet up in the sky, covered in incredible carvings, and all fitting perfectly, in an attempt to keep them up for ever. That's why I don't get much pleasure demolishing a chimney, because I think of those fellows manhandling tens of thousands of bricks, and tons of mortar, to build the thing in the first place.' When he was called upon to demolish a chimney, Fred preferred an old-fashioned method. First he cut a mouth out in the bottom of the chimney and propped up the opening with bits of telegraph poles and wooden chocks. Once enough of the chimney was propped up on wood he lit a large fire, which burned the chocks away so that the chimney would fall in the designated direction.

Harry Forshaw, a demolition contractor and a long-term friend of Fred's, gave him a lot of his felling work. 'He did our first chimney in 1971 and that was the first of many he did for us. Some of them were in very dodgy spots, but he always managed to get them down in the right place. I preferred to employ Fred because he did it in the old Victorian way of burning the chimney down. I thought it was a safer way than the whole dynamite job.

So that's why we always used him. That and the fact that he was such a character and it was a joy to watch how he tackled each job.'

Although Fred would have preferred to be climbing the chimneys to repair them, if they had to go he took great pride in the precise and dramatic way he went about the business. 'My main competition is the dynamite men, when it comes to felling a chimney. They'll come along and blow it up in half a day so they only need a quarter of the money that I want. None of those chimneys are done with the pit props and the big fire, like they did it in 1899. But I like doing it that way because it's more spectacular and you get a bit of a build-up to it, whereas when you press that plunger, boom, that's it, it's over with. But when you press that plunger you've not really done so much; you've just destroyed something that took a few men a long time to erect and a lot of hard bloody sweat and labour. When they finished it off, no doubt they put the Union Jack up, and you've just blown it up with pressing a button. Whereas if you've got to hack your way through three foot of brickwork at the bottom with a few hundred tons squeezing on it, it's not died too easily, has it?'

Fred's other main competition came from what he regarded as the lunatics in the business. 'I've had instances where people have come to a job where I've wanted a couple of hundred pounds to do it and they've done it for twenty-five quid. They're madmen, no pit props, nothing, just chop the bottom out like a tree, and they don't know to an hour when it's going to go. They just keep bashing at it on each side and when it's creaking and groaning they just run out the road at great speed. You don't need anything, only a sledgehammer and a bit of lunacy, and you're in business. Then you've got the insurance men to contend with. They'll send some little lad out from their office who's never seen a chimney fall down and he's got to work out all the risks involved. By the time it's gone back to HQ it could have been Groucho Marx who's knocking the damn thing down, which upsets me a bit when I think of all the ones I've done that have gone OK. Yet the premiums keep going up and up and up all the time.'

Filming a felling was quite dangerous for the crews. On one occasion Fred was doing a felling at the Lilac and Briar Mill at Shaw, near Oldham, and just one corner of the mill was still standing. As the chimney fell, it would be heading for that corner. Nobody could tell how far the bricks were going to go after the chimney hit the deck. The camera crew were on the top floor of the corner of the mill and as it fell all the bricks from the chimney headed straight for them. On the film it looks as if they're going to go straight through the front of the camera. 'Usually,' Fred recalled, 'they didn't just have one camera. They had four or five cameras in all sorts of exciting positions, like where it really looked as though the bricks were flying past the camera and the next one that were coming would blow the lens in. But it never did because, when you do it with the fire and the pit props, there's not a lot of chance of getting hit with anything if you're one and a half times the height of the chimney away.'

As Fred's fame spread, his chimney-fellings began to attract bigger and bigger crowds. But it was a source of great regret that this fame should rest on his exploits as a feller of chimneys rather than his other work of conservation and restoration. Over the previous decade Fred had flattened much of the industrial landscape that he loved so much. Alan Crompton, like Fred, had found a strange beauty in the old buildings and regretted their passing. 'It was such a gradual thing, losing all the industry in Bolton. You had to learn to adjust to your new environment and get on with your own little hobbies. I had hobbies and Fred had his hobbies. So we just let the world go by. Then all of a sudden you look round and it's changed dramatically and for the worse in a lot of ways. There were no graffiti in the old days. People donkey-stoned* the steps, they donkey-stoned the window sills, they swept outside at the front, they swept the main road. This has all gone now. In that sense it's deteriorated. It's very sad really, for somebody of my age. You've got to turn a blind eye to it or you could get very depressed.'

* Donkey stones were scouring stones, originally used in the textile mills of Lancashire and Yorkshire to create a non-slip surface on greasy stone staircases. Available in three colours – cream, brown and white – they were taken up by housewives, who would stone the edges of their steps after they had scrubbed them. They got them in exchange for old rags from the rag-and-bone man who came round with a donkey cart, shouting, 'Rag, bone.'

The views of Alan Crompton were, and still are, those of a vast silent majority of the British public. They reflected the values of an older generation: hard work, respect, discipline and doing a good day's work for a good day's pay. As Don Haworth made more programmes with him, Fred became the embodiment of these values for many ordinary folk who didn't usually have a voice on television. Fred became a celebrity by standing for everything that TV was not and his notoriety, as he used to call it, gave him a receptive audience for his views. Nowhere were these views on society summed up better than when he was filmed in the graveyard of a church he was working on where many of the gravestones had been vandalized.

'We arrived the other day to have a look at the bell, and the skyline in the cemetery had changed. God, you know, twenty-four gravestones had been pushed over and smashed to pieces. It's about ten thousand pounds' worth of damage. What sort of people are these who can come around and shove over things like that, that have stood here for a hundred years? They're evidently more than children to push a stone over like that. They're out of work and very unhappy and have a grudge against society as we know it, all nice and peaceful and proper, and they just want to disrupt everything they can bloody lay their hands on. Why have we gone like this now? Things were really bad in the 1930s. People were actually hungry then. None of these people are hungry and have no bloody shoes – and yet them, them fellers in the

1930s, they didn't do it, did they? The thing is, they lived in a world of bloody fear; they knew if they did the wrong thing it would be trouble, and now they have no fear of anything. We've definitely gone too bloody soft. *Poor little Johnny couldn't help it; he's under a lot of stress.* I mean the bloody Ayatollah, if you pinch a loaf of bread, bang, it's the whole finger off. If they did a bit of that here, maybe things would be a bit straighter. In the Isle of Man, when they had the birch, vandalism were practically non-existent – and now with the wonderful bloody Common Market they've stopped using it; can't do it no more because the Common Market says so. It'll never get any better until they bloody fire them into line; National Service or something like that. There are all these arguments against it, but the people who argue against it never come up with a bloody solution, do they? Never! Just gets steadily worse day by day, so what can you do? There's a lot to be said for the olden days, I say, and how they did things. They were better craftsmen, better at everything really, and they had more respect for property. Now they have no respect for anything.'

Writing in the *Sunday Telegraph* in February 1999 just before the first series I made with Fred was shown, David Robson summed up his appeal. 'Fred Dibnah, philosopher and steeplejack, has been representing the people for nearly twenty years. Through the magic alchemy of television, he has become the embodiment of old-fashioned British common sense ... His vigorous views on anything and everything have turned Fred into a minor folk hero ... On his own admission, he has "never been over-gifted with brain power". His genius lies in cutting through the stiltedness of so much television documentary with its carefully scripted voice-overs, and speaking to the viewer in simple colloquial English. To hell with the syntax: every word rings true as a bell. It is an enviable knack that surprisingly few have mastered. Naturalness is an elusive commodity.'

Remembering those first films, Fred's friend Alan McEwen said, 'Many things made Fred so popular. For starters he was a bit of a comedian. When you saw him on the television, you saw this diminutive, flat-capped character; a working man who typified the northern mill towns, whether they be Lancashire or the West Riding, up to just a short time ago. You had members of your family like him; you certainly had friends like him. You went into the local pub twenty-five years ago, there was always a guy like Fred in the corner, a flat cap on and probably a dirty face. It could be a builder, it could be one of the old-type plumbers, he could have worked in a foundry, or he could have worked in an engineering factory or a mill. Fred was the epitome of that true grittiness of the north. He put things over in his Lancastrian tones; very easy to understand, pretty blunt, down to earth, and you knew what you saw was what you got.'

When he began to appear on television, Fred's views on Britain's glorious industrial past didn't always find favour. He said he had been born into a world going downhill fast. The high point of civilization had already been reached: that was in 1913, when Britain's coal and cotton production achieved a level never to be equalled. He also caused a storm of protest from viewers by his constant swearing. Even Alan Crompton, one of Fred's best friends, was a bit shocked as it wasn't something he'd been used to hearing from his mate. 'There's one thing that Fred did change dramatically and that was his language. When he was young Fred never swore, but over the years it started getting a bit strong and we used to tell him off about that but it didn't make any difference. I said to him once, "My daughters will be watching this programme. Cut out the swearing, for goodness' sake." And he'd give a talk at our Canal Society and say one or two blasphemous words, so a few Christians who were there walked out. But you see, that was part of his colour.' Derek Roscoe said, 'He used to say, "You wanna hear them at the BBC; they all swear." '

The early eighties were a busy time for Fred. The Lancashire cotton industry had all but disappeared and many of the mills were being demolished, so there was plenty of work for the demolition men, scrap men and a good chimney feller. But Fred still found time for his engines and going off to steam rallies at the weekend. One of the friends Fred met at the rallies was Len Crane and Fred would ask him to some of the fellings. 'We had good times when I was invited to help a little bit when he knocked the chimneys down. That was good fun, terrific fun, and Fred was a very clever man; it was a skilled job the way he used to lay it all out. He knew exactly where the chimney was going to drop. If it was a hundred and twenty feet high, he'd hammer a stake into the ground and say that's exactly where the top of the chimney's going to land. A big chimney like that, fourteen foot square, with all those bricks falling down, and as it went down it used to land exactly on that stake and everybody used to clap.'

Fred's army of fans was growing. Seb Coe, the leading middle-distance runner who went on as Lord Coe to spearhead London's bid for the 2012 Olympics, was one of them. He remembers watching Fred at home with his father, who was also a great fan. 'It was the enthusiasm, the pure unadulterated excitement, the moment that a chimney was about to come down. It really was a work of art to be able to drop those things to within millimetres of where he said it was going to fall. And then it was standing back after the first pint of the day and seeing the pure pleasure that Fred got from doing it.

'During the mid-1980s I was doing some work with the kids at Bolton United Harriers when somebody said to me, "Is there anything you want to do while you're in Bolton?" And I thought yes, I'd actually like to meet Fred Dibnah because, like everybody else, I'd been enchanted by what I'd seen on television. It was required viewing. And about an hour later, the guy came back and said, "We've fixed it up." In a short time I was in Fred's back garden wandering round his workshop, and the en-thusiasm was just unbelievable. I guess I related to it because my father is a mechanical engineer and I know the passion and excitement that he has when wandering round steam engines and lathes and all that sort of stuff, and it was infectious. The guy was very funny and wonderfully politically incorrect. In fact I'd say Fred was the father of political incorrectness. And I just thought he was enchanting. The television programmes introduced a generation of youngsters to practical engineering with a touch of style.'

But danger was never far away and it was probably that element of risk that attracted so many viewers. Eddy Chatwood, Fred's steeplejack friend, worked with him on many of the fellings. 'We used to get many near do's, but the one that sticks most in my mind is one we were doing at Cockermouth. The night before we were due to get it down, someone lit the chimney in the middle of the night. By the time we got there, it was difficult to save it. It were crumbling away. Once we'd got the fire out the props were badly burned and the chimney was only just about standing, so we had to go back underneath it to get some new props in that would hold it up till the morning. I don't want another one of those; that was really frightening.'

By now old friends like Michael Webber were beginning to find all the celebrity business rather odd. 'Before he was a celebrity you could be friendlier with him, and then when he became a celebrity you had to share him. I always remember being on a chimney with him after his first series, and there was a school field at the bottom of the chimney. All the kids were shouting up to him, "Right Said Fred", like the Bernard Cribbins song, and it just seemed very strange! And it was strange for Fred as well. But I think he loved it. He must have loved it. He loved being the centre of attention.'

Through the programmes Fred's fame as a feller of chimneys continued to grow. But at the same time as we were watching Fred change the industrial landscape of northern England in Don Haworth's films, we were watching greater and more far-reaching changes to the whole industrial landscape of Britain on the news. The eighties were the Thatcher years and Fred's chimney demolitions became a metaphor for the wholesale dismantling of British industry that Margaret Thatcher was responsible for. We would watch Fred flattening the industrial landscape that he loved so much and then on the news we would hear of more steelworks closures and follow the death throes of the once great mining industry through the bitter, long-drawn-out miners' strike.

The profound changes Margaret Thatcher had set in motion as Prime Minister altered much of the economic and cultural landscape of the United Kingdom. She curtailed the power of the trade unions, cut back the role of the state in business, and dramatically expanded home ownership, all of which was intended to create a more entrepreneurial culture. She also aimed to cut back the welfare state and foster a more flexible labour market, which she believed would create jobs and help companies adapt to changing market conditions. Exacerbated by the global recession of the early 1980s, her policies initially caused large-scale unemployment, especially in the industrial heartlands of northern England and the coalfields of South Wales, and accelerated the de-industrialization of Britain.

The industrial landscape was changing for ever. It meant there was plenty of work for Fred and plenty of fellings to be filmed. The recession of the early 1980s saw manufacturing, the main area of union strength, shrink by half while unemployment soared. Union membership plummeted from a peak of 12 million in the late 1970s to almost half that by the late 1980s. By stripping the unions of their legal protection Margaret Thatcher diminished their power. Flying pickets, the shock troops of industrial warfare, were banned and could no longer blockade factories, ports and public bodies during disputes. Strike ballots became compulsory. The closed shop, which forced people to join a union if they were seeking employment in a particular trade, was outlawed.

The miners faced brutal attacks by the police, who utilized techniques of suppression never seen before in mainland Britain. Mounted officers charged at pickets and through the streets of mining communities. A national task force was created of heavily armoured riot police, which was used to mount military-style attacks. Miners were prevented from freely moving around the country, and special courts were created to deal with the large numbers of arrests made. During the strike many pits permanently lost their customers, leading to the closure of pits on economic grounds and the destruction of many mining communities.

As we watched Fred's beloved chimneys coming down, the number of people out of work in Britain rose above three million for the first time since the 1930s. The official jobless total, announced on 26 January 1982, was 3,070,621, which meant one in eight people were out of work. Labour leader Michael Foot was reported to have said, 'When Mrs Thatcher came into office there were five people chasing each job and that was bad enough. Today there are thirty-two people chasing every vacancy and in some parts of the country it's double that.' And Tony Benn, MP, said, 'She destroyed our manufacturing industry. She brought unemployment to the highest rate we'd had since the pre-war years. She began to dismantle the welfare state. She deprived the health service of the resources that it needed. She deprived education, schools and so on, and she did enormous damage to the fabric of society.'

But the destruction of manufacturing was keeping Fred busy and providing plenty of exciting material for the film makers. By now, though, Fred and Alison were beginning to discover the downside to being in the public eye. In their years together after the filming had started, they never quite managed to come to terms with fame. 'This celebrity business,' said Fred, 'is all very well if you could just escape from it every now and again. Two or three years ago I could go in my shed in the back garden after tea and start work on my steam engine, and know full well that I wouldn't have any phone calls. Nobody would bother me. Now it isn't like that. I just get outside and Alison's at the back door, "Phone for you," so I've got to traipse across and answer the bloody phone, to find that somebody wants me to go and do an after-dinner speaking thing. And you know, I get very little money out of it. In one instance I got two sacks full of rivets for my new boiler as payment for an after-dinner speech at a Rotarians do, and I get the odd bottle of wine or twenty quid here and forty there. But I'm fast looking at all that business and thinking really I'd be better off without it. I'll be better off concentrating on my back-yard operations.'

Another obstacle arose that looked as though it might affect Fred's progress on the traction engine. Alison wanted a holiday. The last they'd had, steam rallies apart, was their honeymoon. Now, with the oldest of their three children in her teens, Alison felt the time was right for another vacation. So Fred found himself at Blackpool, and under orders not to pull a face about it. It didn't suit Fred. 'Really, I get most of my pleasure out of working. It's nice to see the kids all happy, but when I was little I never liked going away. I used to have little tantrums and they'd throw me on the coach for Blackpool, which I never liked. Perhaps that's why I don't like it now. It was done to me when I was quite small, you see, the damage, as you might say.'

But Alan Crompton remembers the real reason for the holiday. 'He had to work. The only time he went on holiday to Blackpool, it were to do a job at a chimney while Alison played on the beach with the children. The holiday had come out of a deal with the owner of an engineering works who had a small chimney that needed felling. Fred needed a new front plate for the boiler of the traction engine so he agreed with the proprietor of the works that he would knock his chimney down if he made the plate for him.' The 'holiday' was captured on film by Don Haworth. On the first day Fred went to the beach with the family and everybody played happily. But after a couple of days the weather changed and play was rained off. The family piled into the Land-Rover and Fred drove them to his chimney. The scene in which the children sat in the Land-Rover watching as their mother, anorak hood tied around her face, struggled in the rain to help Fred ladder the chimney spoke volumes about his views on marriage. But strangely it just added to his appeal as more and more people switched on for the real-life soap opera as Fred's relationships came into the public domain.

'He just wouldn't have a holiday,' Alan Crompton said. 'It were a waste of time, you see. Holidays were just for the rich and the idle, not for Fred. I think the nearest he came to it apart from Blackpool was one time when Alison said to my wife, "I think Fred will go away with Alan," because we were quite close. She started making tentative arrangements with us and she were thrilled to bits but something happened. A job came up and she said, "I'm sorry, we're not going." That were the nearest they came. Fred never had a holiday. He wouldn't know what to do at the seaside or in the countryside. He'd be lost.

He'd rather be in the backstreets of Birmingham looking at derelict warehouses and old engineering shops!'

But Alison wanted a proper holiday. The pleasures of steam rallies were beginning to wear thin and, as the years went by, her thoughts turned increasingly to sunnier vacations, beyond the range of Fred's roller. Every year she said she wanted a holiday. Other families went on holiday so why couldn't they? What Alison had in mind was a trip abroad, to Greece or somewhere similar, with sunshine and blue skies. And that was the snag: Fred had been abroad once when he was in the army, and he didn't like it. 'Well, I think in Greece really there's nothing for a man like me. Lying there on the bloody sand all day, watching the sun shine, it isn't really my style.' So if Fred wouldn't go, Alison would have to take the girls on her own. She booked a holiday to Greece and then told Fred. When he started complaining, Alison paid for the holiday. Fred got very upset about it; but Alison had waited eighteen years for this trip and she was determined to go. Soon after they got back there was a bombshell for Fred. Alison wanted a divorce. She'd had enough of the world of steam engines and factory chimneys and wanted some normality in her life.

'I don't know why,' said Fred, 'because the bloody chimneys had kept us alive for eighteen years. And life had got to a stage where I could semi-retire. Then all of a sudden it looked as though I'd have to get a mortgage and work for another eighteen years to buy Alison a house and fix her up in the style that she's become accustomed to.'

Fred's brother, Graham, sees it differently. 'Alison didn't like the steeplejacking and steam engines too much – like a lot of people wouldn't. It's all right at the beginning, but then after a while it wears off you. I think that's why they ended up parted, and later divorced. I reckon, if he'd have spent more time with her, and took them on holiday now and again, instead of being always tied up with the engines and his work – they may have still been together.' Many of Fred's friends agreed. Neil Carney said, 'I think that marriage terminated purely because of the time he spent rebuilding his roller. He used to work till way past one o'clock in the morning sometimes, down in the shed. He'd much prefer to be with his roller than with his wife really.'

When Alison left Fred began to muster his assets, including, regretfully, his 1927 AJS motorbike. 'I always knew it was worth a bob or two, and now that I'm a bit desperate for money it looks as though it's got to go. It's a great pity really. I suppose it can go for auction at Sotheby's and then we might get two lunatics bidding against each other and we might make a couple of grand. It's a shame really because it's one of the few things I've had to part with that I really liked.' The bike went for auction and Fred got £1,700 for it.

After Alison and the girls had moved out, Fred came to see the stress of the celebrity business as one cause of their domestic trouble. 'Sometimes the bloody pressure of it all gets a bit too much. I don't think Alison was a person who liked excessive pressure. She were more the person who were quite content to sit there watching telly and just go out and do the things she wanted to do, without having all this extra load thrust on her back. There is no doubt whatsoever, she were a good help to me and I'm going to miss her. It was just that all the pressure got too much for her.'

Fred was on his own for the first time and he was lonely. He'd lived at home with his mother and father till he was twenty-nine and ever since then he'd been with Alison. 'Up until the time that Alison left,' he said, 'I'd always thought of myself as a bit of a loner. But when I was left to my own devices I soon realized that I wasn't. Even when you've fallen out with someone there's a sort of companionship in just having them around the place. But without Alison and the girls the place felt totally empty.' Fred began to realize that he'd not devoted enough time to his wife and his daughters but it was too late now. 'The trouble was,' he said, 'I'd got too tied up in my own things and neglected my family a bit. I should have taken them on a few holidays, but for me steam rallies were the best holidays you could ever have. Sitting around on beaches was a waste of time for me, but I suppose sitting in a living van in a wet field full of steam engines wasn't always their idea of having a good time.'

Fred had now got all the time he wanted to play with his engine, but he was finding he couldn't settle to anything and he didn't know how best to occupy himself. 'I began to hate the thought of getting home from work or coming in from the pub,' he told me later. 'I found it all very quiet. The only good thing about it was that I didn't have the kids playing records and there was nobody there to shout at me and tell me what to do. I didn't like it but perhaps it was all for the best. Alison wanted a more modern sort of life but I've never really agreed with this equal rights thing so I don't suppose there was any way I was going to change from being the Victorian-style husband everybody said I was.'

But Fred knew that hadn't been the only problem. Everybody who knew Alison said she was a very private person, but once Fred had allowed television into his life everything connected with it, including his marriage, had become public property. 'When television came along and made me notorious,' he said, 'all sorts of people started ringing me up to do this, that and the other. But Alison wanted to live a quiet life and the pressure of it all got too much for her.' In later years, when I was working with Fred, the first question the majority of people would ask him when we were out filming would not be about his chimneys or his steam engines, it would be about his wives. Today, more than ever, people want their five minutes of fame. Fred had a lot more than five minutes, but there was a price to pay for it.

With Alison and the girls gone, Fred passed the long and lonely evenings in the steam workshop he'd built in his back garden. 'I'm managing all right. The cooking's a bit of a mess because you've got to stop whatever you're doing to go in and cook your dinner and it takes a long time. I'm used to it being ready on the table for me and just racing in and going at it. I did my first trip to the supermarket the other day, shoving the trolley around. I felt a right berk – never done anything like that in my life before. The thing is, it's very peaceful here now. I've got rid of all that bloody pop music and you've a bit more time to think. And now all of them have gone it will give me a chance to lick the place into shape. I've sanded the floor in the new extension and now I can play my Johann Strauss records.'

Looking back on it years later, Fred reckoned the time just after his first divorce was one of the worst in his life. 'Living here all alone weren't much fun really. I'd come home, I'd put my pie in the oven and put the kettle on, and watch a bit of the news and try and keep the time for going to the pub as late as possible. More often than not it ended up somewhere round about eight o'clock that I'd roll down to the pub and join the other divorced characters in a similar situation. I soon realized that you couldn't go on like that for ever or you would become just another bloke who props the bar up down the pub with not a lot to show for your life here on this planet.'

Then along came Sue.

6

Preserving the Past

Sue Lorenz was a social worker who took an interest in Fred's engines and his back-yard steam workshop. She first met Fred at a big steam rally in Cheshire. It was in the middle of all his divorce problems and Fred was very drunk. In one of Don Haworth's programmes Sue recalled that first meeting. 'He looked awful. He was sad and miserable, quite a pathetic sort of figure, with none of the bounce that people knew him for. He invited me to a chimney-felling in Oldham, and he nearly brought it down on the top of all of us.' It was not the most auspicious start to a relationship, but shortly after that Fred rang Sue and asked her if she would like to go with him to a talk he was giving to a group of fans in Halifax who called themselves the Fred Dibnah Appreciation Society. She accepted, and recalled that when they arrived at the little pub in Halifax, there were lots of men similar to Fred, all wearing flat caps and smoking Woodbines.

Fred's Appreciation Society had provided a room and the get-together turned into one of those nights that Fred talked about for years afterwards. 'It was a scruffy little place and everybody was hopelessly drunk at the end of the evening. I remember waking up in the middle of the night to a terrible row and the landlord and the landlady were throwing things at each other. I got up and could see the projectiles going through the windows downstairs. It sounded like there was a bloody war on. When we got up, all the windows were broken down below. I'll never forget that as long as I live.'

Nevertheless, Fred married Sue and in one of the programmes she described what her new husband was like at that time. 'Fred had some really bad traits when I first met him; he was terribly self-centred and thoughtless. A silly example is, he would go to the pub and he would be with a group of people, and he wouldn't buy a round. He was so busy talking about himself that he wouldn't get his money out, and I used to feel so angry with him. The other things that really upset me, because I could see that they were damaging him, were his bad habits of drinking and smoking. I eventually persuaded him to cut down on his smoking, and then of his own free will Fred said, "I'm going to give up." And to encourage him and support him, I bought him presents for weeks, every day that he didn't smoke.'

By this time there were not many chimneys left around Bolton so Fred was having to go further afield for his work. Very early in his steeplejacking career Fred had done some work for the Revd Barry Newth when he was vicar of Clifton, near Bolton. He'd struck up a friendship with Fred and recognized the early signs of a unique artist and craftsman. In 1986 Barry Newth found himself vicar of Kirkby in Malhamdale, in the Yorkshire Dales. When he took up his post he discovered that the great medieval church, which was known as the Cathedral of the Dales, had neither a lightning conductor nor a usable flagpole. There was only one man for the job. Fred went to the Dales with his old steeplejacking mate, Eddy Chatwood. While they were doing the lightning conductor they had to dig down into the ground by the side of the tower for the earthing strip and to their surprise they dug up somebody's bones. 'So we had a reinterment ceremony,' said Fred. 'I don't know who he was, but Oliver Cromwell had been there so it's a pretty old church.'

For Eddy, working in the countryside made a welcome change and he remembers it for reasons that were not directly connected with the steeplejacking. 'We were in Kirkby Malham on the church,' he recalls, 'and it's the only time I've got a vicar drunk. When we discovered he were on whisky and soda, I told the barmaid to put three in. The lass behind the bar did this for about three nights, and Fred said, "I don't think his wife likes us, her scowls are going greater." So the vicar didn't come to the pub with us any longer but he did get us to give him a lift to remove sheep from one church to another, where they kept the church grass tidy. Now Fred and me had never thought we'd end up running down a country lane behind a load of sheep. Fred had a genuine shepherd's crook in his hand, and while running along this road, a lady recognized Fred and came from her bungalow and said, "Oh! I know who you are. But I didn't know you were a farmer." '

In 1987 Fred and Sue's first son was born. They named him Jack. Fred was still working in Kirkby Malham at the time, and while they were there Barry Newth asked him if he could make a weathercock for the church. 'OK,' said Fred, 'I'll do you a deal. I'll make you a weathercock if you'll christen my little lad in your eight-hundred-year-old font.' So Jack was christened in the parish church of Kirkby Malham.

Fred was very proud of little Jack. He used to say, 'He turned out to be quite an exciting little fellow who liked more or less everything I did.' Jack went to his first chimney-felling when he was about eighteen months old. He watched the tall stack fall down and as it hit the ground he said, 'It's all gone,' and started crying. They went back to the house and somebody had a video camera so they watched the felling on the television. Jack was upset at seeing it fall again and Fred was so proud of him, saying, 'He's just like me, he hates destruction.'

Apart from his work on chimneys and spires, Fred had devoted his life to steam. Over a period of around twenty-five years, he'd rebuilt one steam engine and got well on the way with the restoration of another. Now he'd reached fifty and he'd achieved the ambition he had cherished since he bought his first engine, a great steam-driven workshop in his back garden. Put together from scrap and the surviving parts of ancient machines, the workshop now had the capacity to restore many more steam engines, and to take Fred into a new line of business – steam preservation.

At the beginning of the twentieth century there were steam engines everywhere in Lancashire, busy providing power for the county's cotton mills and factories. Steam had played a major part in the region's industrial history, but during the time that Fred had been doing his steeple-jacking thousands of these engines had been destroyed, along with the mills, pits and forges that had housed them. The scrap merchants had become wealthy as they stripped the brass and anything else that was worth salvaging from the engines. Fortunately, dedicated bands of enthusiasts like the Northern Mill Engine Society began to appear; people who were aware that, if action were not taken quickly, all of our industrial heritage would disappear. Fred recognized that, while the steeple-jacking work was going into decline, there would be new opportunities for him in the restoration of steam engines.

John Yates is an inspector of historic buildings for English Heritage. He said, 'Fred was in there at a time when the first generation of industry and old machinery enthusiasts were about; when people were just beginning to get the idea of rescuing traction engines from scrap yards and raising historic boats and repairing mill engines and pit winding engines. He was in right at the beginning of that and those wide-eyes loonies will have influenced him, because he's one of them.'

Fred's old friend Alan Crompton was already involved in the restoration business and he began to include Fred in his projects. 'At that time I were doing some restoration work for Manchester Museum and I wanted some small brass castings. I had to form a piece of sheet metal round the brass casting and I couldn't do it, but Fred could. So he did one or two of them for me and then I did some things for him. That was always the way Fred liked to do things, with no money changing hands. I made him a full set of wooden patterns, because I had a wood lathe. I also turned him some finials on his sheds because I were into wood-turning at that time. Then I did a full restoration job on his grandfather clock.'

The first big renovation job that Fred got himself was in Wales. One day someone from Caernarfon Council called about a chimney they wanted Fred to have a look at. At first Fred wasn't interested. Caernarfon was a long way from Bolton, the chimney was only about seventy-five feet high and Fred knew he wouldn't be able to make much money on it. But then the man on the phone said, 'We've got a steam engine here as well.' When Fred heard that he decided to go and have a look.

The chimney and the engine were at a place called Glynllifon Park, a few miles outside Caernarfon. It was a big estate with a great house and a workshop where all the maintenance work was done. 'When we arrived,' recalled Fred, 'the chimney stack was like one you'd get on a typical Cornish pumping-engine house. It had ivy round it from the roof right up to the top so you could hardly see any of it. I couldn't tell how much it would cost to renovate that chimney because it was completely covered in all this ivy. You could hardly see the brickwork and I knew there might have been all sorts of cracks and holes behind the ivy. The man from the council said, "It's only a feasibility study that we are doing. Just give us a price for coming over and taking the ivy off and then you can report on its condition and give us a price for doing it up." So I said OK to that.'

It was the first time Fred had ever had to leave home for a job. 'We'd never camped out or been in lodgings before. So we went and stayed in the local pub down the road just like travelling tradesmen.' The first day on site Fred put the ladders up, got all the ivy hacked off and found that the chimney needed pointing right from the top down to the roof. There was no lightning conductor on it either and it wanted a couple of iron bands round the top as well. In the engine room there was a beautiful single-cylinder 1854-vintage stationary steam engine with a ten-foot flywheel that had driven a big frame saw. Unfortunately there wasn't much left of the saw. The engine itself was in a very sad state because it had been robbed of all its brass. Immediately above it the skylight was broken and the rain had been falling on to the engine for years.

Fred gave the council a price for doing up the chimney

and for making the engine run again, but unfortunately he only got the contract for renovating the chimney, which wasn't the one he really wanted. When he came to do the job, the weather was fine and progress was good. But one day it rained and they couldn't work, so Fred got the key to the engine house so he could have a look at the steam engine. He found that, apart from the fact that all the brass had been stripped off, the engine wasn't in bad condition. The only trouble was that it was half buried under timber, wooden posts and fallen plaster. Once he'd seen it, he really wanted to have a go at fixing it.

Some time later, after the chimney was finished, Fred got a job in Deganwy, only a few miles down the coast. While he was there he met somebody from Caernarfon County Council. 'What about that steam engine?' Fred asked him. 'Have you done owt about it yet?'

'Oh, we'll send you a tender form,' the man said.

So Fred put a tender in and got the job. He went over to Caernarfon, where he pulled the engine to bits. Then he took all the parts back to Bolton and spent six months working on it.

By this time Fred had a new assistant. Neil Carney worked with Fred from 1989 until 1994. He had known Fred for a long time. He'd worked as an engineer for Greater Manchester Transport and had first met Fred in the mid-sixties at a Burtonwood Traction Engine Rally, just after Fred had bought his roller. He kept in touch but didn't see a great deal of Fred until he took early retirement from his job as depot engineer at the Wigan garage.

By this time he was living in Bolton, so he went round to see Fred. As a retired engineer he'd got the mechanical skills that Fred needed for the Glynllifon restoration job, so Fred asked if he'd be interested in working with him. Neil took the job on and helped Fred restore all the parts, take them back to Caernarfon and begin rebuilding the engine. While they were away, Fred's second son, Roger, was born.

With Neil, Fred worked on the Glynllifon engine, on and off, for over six months. When they'd assembled it, they hired a big compressor with a great wagon engine in it. They got the steam engine going on compressed air because it still didn't have a boiler to give it any power. Part of Fred's contract was to jack up the boiler and get a boiler inspector to look at it. But it didn't need an inspector. It was in such bad shape you could have put your boot through it. So what were they to do? It looked as though they could only do a cosmetic job – paint it matt black, put all the brasses back and the pressure gauge and paint 'Thomas D. Winton, 1854' in cast-iron lettering over the firehole door. And that was what they planned to do until they met a man in the local pub one night who said he knew where they could get hold of a boiler. 'Mr Thomas, the pork pie manufacturer down the road, has just done away with his boiler. It's a good 'un and it's just laying on its side in his yard.'

One evening when Fred and Neil had finished work on the engine, they decided to have a look at this boiler. It had twelve cross tubes, which meant it was a very good steamer. Fred let the council know about it and got another £7,000 for installing the boiler. They lagged it with mahogany and put brass bands round it and Fred thought it was one of the nicest things he'd ever done. It was such a good job in fact that he won a £500 prize for the year's best stationary engine preservation effort.

Alan McEwen, a friend of Fred's and himself a boiler-maker, had a great respect for Fred's skills. 'Fred had a natural ability as a boilermaker. He was an observer. If he had a particular interest in something he'd look about him, and he could master anything. He studied it for him-self and he had a hands-on approach of experimenting. Eventually, over the years, he developed his own method of carrying out boiler work to absolute super 100 per cent worthiness of the craft. Many people have asked me what his boiler-making was like. "Absolutely superb," I always said.'

During the Glynllifon job Fred had come to appreciate the engineering skills that Neil Carney was bringing to his new venture. They'd worked well together, so Fred asked Neil if he'd be interested in staying on to help him with the steeplejacking side of the business. Neil was happy to accept on condition that he wouldn't have to do any climbing.

'Steeplejacking was a job I enjoyed,' said Neil. 'There was a technique to laddering and I soon learned that. I never needed to go up because I was the groundman. In the earlier days he used to have young lads working with him and if he wanted anything he'd shout down, but they'd never be there. They'd be chatting up the local talent instead of paying attention to the job. So, he said, you want an older person down there really, who's willing to keep an eye on what's going on.

'A lot of Fred's chimney work involved pointing and for this he used a special mastic that he mixed himself. He'd learned about it from Faulkner's, the Manchester steeplejacks who had been his great mentors. Most builders and property repairers just use sand and cement, but mastic has got some white and red lead in it mixed with linseed oil. Fred had got the formula from one of Faulkner's men and it was pinned up on the back of the shed door. It really was good stuff; mastic is always better than sand and cement.

'Occasionally steeplejacking involves some yard work. For example, if we were making bands to put round a chimney, they would have to be worked out and made to length, then drilled and bent at the end where the bolts go through. That was usually my job, working the length of them out. Although Fred was an excellent craftsman and he could achieve the desired result by one means or another, he never claimed to be much of a mathematician. "That's one for you," he would say. "You work that out." And, of course, I was quite happy to work out the length of these bands and mark them where the bends were to be. Then Fred stepped in and we got them hot in the fire and he even had a little pair of rollers and we'd get steam up from the boiler. I'd be up early in the morning to light it. We'd have steam up by ten o'clock or whatever and then we'd literally roll the bands through these rollers to curve them to the curve of the chimney.

'With that method of working, chimneys don't fall down. As far as Fred was concerned, it was up to the customer to place the order for the job before the chimney got dangerously close to falling down. That would give him plenty of time to finish the job. He always did finish, even though the customer might have asked once or twice over the months that it took, when are you going to get the job done? But at the end of it they were always well satisfied with the standard of work and they'd recommend him to everybody else. Provided you were not in a hurry, Fred always did a very thorough job.

'He was great to work with; we got on fine because we shared common interests. I appreciated the old-fashioned way of doing things because my father was also in engineering and he dated back to the time when they did things that way. When pottering in the yard or working away on steeplejacking we never had any arguments. We'd consult one another on the best way of doing things.

'Fred was extremely knowledgeable about the old way of engineering. He had a fairly extensive library of books published round about 1900 or even earlier. He did a lot of reading, ten times as much reading as writing. He was never a great one for writing, apart from his signature on his books which he always wrote in his nice copperplate. As regards letters to customers or replying to their letters, he would always get his wife to do them. He was an excellent draughtsman because he was art school trained. He did some beautiful illustrative drawings of the way a job should be done. If a customer was in doubt as to how he was going to go about it, next time he'd take him a drawing of how he intended to proceed.'

The birth of Roger had let Fred in for another year or two of morning marches to the nursery. But with two little sons he was so proud of, changes were seen in Fred. 'It didn't bother me at all,' he said, 'shoving this guy about. He was ever such a pleasant little chap and he never complained about anything. The only time he ever cried was when he was hungry, which was a natural thing to do. Roger was very similar to what everybody says I was when I was little; a pleasant little fellow. Sometimes when I was taking the lads out I'd be taken as the grand-dad. That used to be a bit embarrassing, when people asked me if they were my grandchildren. But I just used to say, "Are they buggery, they're mine. I'm still in good condition, still of breedable quality you know."'

If ever anybody was born to be an engineer it was Jack and, as he got a bit bigger, he started to toddle round Fred's workshop. In those early days, though, he could be more of a hindrance to Fred than a help. 'Jack's a damned nuisance round here,' Fred said in one of the films from that time. 'In his way he's trying to give me a lift, but of course, if I put the spanner down, it disappears and all sorts of weird things happen. You know he's had it. But he's got such a good memory that if you say, "Where's that big lump of iron with three holes in?" he'll go straight to it. He has one bad fault though. He likes digging holes, especially when it's raining, and there's bloody holes everywhere all full of water. I think he's a frustrated coal miner or something.'

Family life was looking up for Fred and he started to show some signs of domesticity. In one of the films Sue said, 'He likes hoovering and if I ask him to hoover, he makes a super job of it. He does it much better than I do. He gets under all the corners because he's a perfectionist. And all the brasswork is his responsibility. He does all that and the window-cleaning. I never ever clean windows because Fred does them all; he loves twinkly windows. I wouldn't ask him to change the baby's nappy, though, because poor Roger wouldn't like to be subjected to being stood on his head, which I think is how Fred would tackle the job.'

And Sue began to have an effect on Fred's appearance. 'Some things have changed a bit,' said Fred. 'I've got more of an upmarket wardrobe now than I used to have. Instead of one suit with a waistcoat and various pairs of jeans, I've now got a grand selection of pullovers and cardigans and two pairs of shoes. And various caps for doing various things in – appearing in public in posh caps; going to work in dirty caps. I have a wash more now than I used to. I have a shower every night. But I still like my oily cap. It has a smell all of its own. I've been in various pubs, and people have said to me, "What's the flashpoint of your cap?" I know that when I'm doing a bit of blacksmithing, occasionally I take it off, and I get hold of some hot iron and it sets on fire, there's so much oil in it. On two or three occasions I've been so inebriated after days out with this load of old iron in the yard that I've actually ended up in bed with the cap on.'

'The load of old iron in the yard' as Fred called it was, by now, being seen more and more on television. Ian Howard, who runs a steam engineering business in Derbyshire, recalls the interest that the programmes were raising in the steam preservation movement. 'When Fred first went on television with his steeplejacking he brought the steamroller into it and there was an instant massive following at the steam rallies! The roller definitely brought the steam engine and the steam rallies to the forefront of popular culture. A lot of people went to the rallies in the hope of seeing Fred, and even though Fred couldn't be there all the time the attendance numbers still went up. Without people supporting the steam rallies I'm afraid they would die a death, so everybody was very grateful for Fred going to rallies and making people aware of what was happening.

'I met Fred initially through these steam rallies. I'd watched his programmes and I could never understand how he laddered a chimney. So Fred sat down in the beer tent one Saturday night and did one of his drawings to show me how he did it. He took the time to explain it all to me and I'll never ever forget that. He was very well regarded in the world of steam. Most of the engine men like a pint and a story in the beer tent at night and Fred would always have a story to tell you. He'd always have a crowd round him.'

Jack Meaker, Ian Howard's partner at Alton Engineering, believes that Fred raised the awareness of steam to great heights. 'Before Fred appeared on television the steam engine world was very close knit. Everybody in that world knew what was going on but nobody outside that world did. Then Fred appeared on the television with his steamroller and it boosted everybody's interest in steam. Fred was a good lad; he fitted in well, and we had some very good nights with him in the pub.'

With his fame spreading and a demand growing from organizers of steam rallies to have Fred in attendance, he was having to go further and further afield. The days of getting up early and steaming all day to a local rally were gone and he now had to arrange to have his roller transported. One of the people who helped him with this was his friend Len Crane. 'I first met him at Ashton Park Steam Rally and we became friends straight away. I had a low-loader for transporting locomotives about and I finished up moving Fred quite a lot all over the country. I even took him to Ireland and that was a bit of a laugh because everybody knew from the television programmes that he liked Guinness, so everybody who came to the rally brought a bag of Guinness. We ended up with what seemed like three tons of Guinness bottles all piled up around his caravan.'

Fred was building up a wide circle of friends in the steam movement. Dick Ransome, an engineer and steam enthusiast from Torver, near Coniston, in the Lake

District, remembers particularly the long telephone conversations he had with Fred on all things relating to steam. 'We used to have marathon phone calls, never under an hour, sometimes as long as two or more. Many a time my wife used to bring my meal through when I was on the phone and I'd say to Fred, "My meal's on the table," but he'd just carry on and then I'd say, "The potatoes have stopped steaming, Fred," and the conversation would still go on. He used to tell me about all the little problems he'd encountered between his last call and the present one and how he'd addressed the problem and overcome it. And at the end of the day, when I put the phone down, my arm would be numb from holding the receiver to my ear for so long. Then when he used to come here, I've got a lot of happy memories of sitting at home in the early hours of the morning, in front of the fire with Fred with his can of Guinness and maybe a Scotch for me. I got my steamroller just after Fred got his and over the years we compared notes on what I've done and what he's done and I learned a heck of a lot off Fred. If I had a problem I'd just pick up the phone and have a chat with him and he was always willing to advise or make a helpful suggestion and nine times out of ten he was absolutely spot on.'

By the mid-1990s Fred's business was changing. Glen Laws is chief executive of Ironbridge Gorge Museums Trust, Britain's most important industrial heritage site, and he summarized the changes to the industrial landscape that had taken place during Fred's lifetime.

'Fred lived through an age in which we started off by knocking down a large part of the Victorian heritage and then grew to value it. He was obviously engaged on both sides of the argument because he was involved with knocking down the great mills. At the same time, he was also involved with the lovingly restored steam machinery. He helped bring it to a wider public and that sort of enthusiasm was very well portrayed in the various television programmes that he did.'

'The trouble with chimneys in those days,' recalled Fred some time later, 'is that not only were they few and far between but also, as I was approaching sixty, those that were left took a lot more climbing. I was getting too old for it. People kept asking me, "How long are you going to keep climbing up chimneys?" But whenever I could get the work I had to do it because, ever since Alison had departed, I'd got a big mortgage. The trouble was that the income from the steeplejacking business had been going down steadily for several years. There were fewer jobs around and a lot more steeplejacking companies to compete with. Sometimes we banked only two or three thousand pounds over a whole winter.'

When he did get the odd call for steeplejacking, it wasn't always what Fred regarded as a proper job. Once he was asked to ladder the India Mills chimney at Darwen, near Blackburn, so that somebody could abseil down it to raise money for charity. Fred found it all a bit strange. 'This guy insisted on having a rope tied round him as he was climbing to the top of the ladders because he said he wasn't very happy on the ladders. But once he'd got off the ladders and he was dangling in the air on the end of his rope, he was perfectly happy.'

Some of his other steeplejacking jobs came from the industrial heritage world, when work was needed on chimneys at mills and factories and pumping stations that were being preserved. And some of these, like the job he got at the Victorian refuse destruction site and sewage pumping station in Cambridge, were taking Fred much further afield. It was many years since the station had been used but the site had been preserved and meticulously maintained by a band of volunteers. Fred enjoyed meeting them. 'They all lived in a romantic land of long ago. They were just like little lads who never grew up. I can't complain because I'm one myself. Really they and a lot of people like them in the preservation movement are romantics, trying to escape from modern life. The pumping engine there had clearly had a lot of time spent on it to make it look beautiful again. It must have looked magnificent when they first completed it.'

At that time, all the travelling that he was having to do to get work was very new for Fred. He'd always worked on his own doorstep around Bolton, without all the extra expense of paying for lodgings when he was away from home. 'Every time somebody opens their front door it's a hundred quid. I remember at one time I used to get very overawed; I didn't really like hotels. The communal eating part of it was always a bit frightening, not picking the right knife and fork up and being followed by strange eyes. At the beginning, with the euphoria and the excitement of a new job far from home it was all right, and then you begin to miss being at home. It's a bit like being a sailor, I suppose, where you get on a ship and you disappear for six months. It's a strange life, when you've been used to coming home for your tea at five o'clock every night for years and years. I got more and more frustrated with the fact that I couldn't get on with any work on my tractor. When I was working at home, even if I only did half an hour in the shed after tea it was a little bit nearer. But when I'd been away all week and came home, I was a bit reluctant to zoom off into the shed in case I got a good talking to. A bit of a shouting at off Susie. "You've been away all week and now you've disappeared off in the bloody shed." '

But even longer periods away from home were to follow. Fred got a letter from *Old Glory*, a magazine for people interested in steam engines and other vintage machinery. The letter said there was a job at Wetheriggs Country Pottery, near Penrith, if Fred could get there quickly enough. In the past a steam engine had worked the mixing gear, the pump mills, the potter's wheels and a blunger, which is a machine that resembles a Kenwood Chef mixer. 'The way it worked,' recalled Fred, 'was that when they set the steam engine in motion the whole thing went round and round and mixed up clay and water to the consistency of milk chocolate.'

Fred won the contract to get the engine, boiler and blunger working again. 'The actual blunger was like a work of art,' he said. 'I really enjoyed restoring that because it were just ready for falling in a heap on the floor, it were so rotten. Mr Dodds, who was the owner of the pottery, said they had been everywhere in the potteries at one time, but this was the only steam-driven blunger left anywhere. Basically, by the time I'd finished with it, the thing was like new, apart from the gearings, shafting and bearings. All the timberwork I did is new.'

Jo Chapman, managing director of Wetheriggs Country Pottery, was there throughout the time that Fred was doing the restoration work. 'Fred worked here in total for six or seven months. The first stage of the job involved the restoration of the blunger and the engine room and the boiler. It was great to have Fred around. We had an agreement with him whereby we could use him for marketing purposes. He didn't mind at all being interviewed by Cumbrian radio and the TV and he would always talk to our visitors. The public just loved to come down and meet him and to see him at work. The job took him a bit longer because of this but Fred just loved to talk to people. One day when I walked into the engine room I could see he was hiding something. He was standing in front of the engine and when he stood to one side there was a brass plaque with my name on it. He'd named the engine after me, so it's called Josephine. I'm very proud of that.'

By this time there was very little chimney work left, so it was a surprise when Fred landed a big restoration job near home, which, tackled a few weeks at a time, would keep him in work for most of the year. Standing 262 feet high, the Barrow Bridge chimney was the biggest left in Bolton. It was the one that Fred had climbed in the dark when he was about seventeen for the ten-bob (50p) bet. Now, in the twilight of his steeplejacking career, forty-odd years later, he'd been asked to repair it. The contract called for pointing the top thirty feet and recapping the top brickwork. To get up there he needed two runs of ladders with eighteen sections in each. He'd been asked by the Royal Society for the Protection of Birds to put up a peregrine falcon's nesting box on the south-west side about thirty feet from the top. Fred wasn't sure about this. Peregrine falcons, he thought, do not like pigeons; they have one for breakfast, one for dinner and one for their tea. The local homing pigeon society had been up in arms about it. They didn't want Fred to put it up. But Fred was never overly keen on pigeons because they'd put him in bed once for a week, with some disease he got off his cap after it had been in a load of pigeon droppings. So he put the nesting box up there.

Not pressed too hard by paid work, Fred spent his time expanding the collection of ancient industrial tackle in his back-yard workshop. But the main focus of his attention was still his steamroller and it accompanied Fred on his increasing number of public appearances. When she was interviewed in one of the last of Don Haworth's programmes, Sue wasn't sure how happy Fred was with the celebrity status the programmes had brought him. 'I think sometimes he struggles with the celebrity aspect of it. If we've got something down in the diary, a public appearance somewhere, he would often prefer to be going off up a church steeple or in the garden doing something for himself. But I think he realizes we've got to do these things to earn ourselves enough money to keep ourselves during the winter.'

Another source of income was beginning to trickle in. Fred had always loved talking and he'd always liked an audience, so when he was invited to do some after-dinner speaking he turned out to be a natural live performer. Many of his first engagements were for groups of engineers. John McDougall, a former president of the Institute of Mechanical Engineers, remembers some of Fred's early after-dinner talks. 'He was always a good talker so we had him at a number of our formal dinners in the north-east of England. The one thing that always comes to my mind when I remember Fred is his flat cap, which he wore even with a dinner jacket. He would be the only one in the room wearing a dinner jacket with a flat cap, which I guess was part of his style. His enthusiasm was infectious. It was very difficult not to sit and listen to Fred when he was telling stories; he was a great story teller.'

Fred himself still found his fame a bit strange. 'It's very odd really, the celebrity business. At the beginning it was quite frightening, eighteen or nineteen years ago. I've got used to it now, and it doesn't bother me as much. What do you do? If somebody waves to you, do you pull your face at them and look miserable and nasty, or do you give them a wave back? If you don't wave you're a miserable bugger, if you wave you're a big-headed bugger. What can you do? You're stuck.'

Fred never changed. Fame never affected him. By the second half of the 1990s he'd been on television for nearly twenty years. After the first documentary Don Haworth had made another eighteen films about Fred, following his life as a steeplejack and steam enthusiast and the attendant strains his passions had put on his marriages. But to his friends and the many people he had got to know in the steam movement he was still the same Fred. Andrew Semple is the president of the National Traction Engine Club. 'He was very much one of us,' he said. 'We all knew him before he was a TV celebrity and he didn't expect to be treated any different to the rest of us. If you were at an event, Fred would get his leg pulled just as hard as anybody else. There were no sides to him. He was always that same jovial character he had always been.'

Jobs like Glynllifon and Wetheriggs were getting Fred more involved with the preservation of the industrial heritage he loved so much. His friend Alan McEwen shared that passion and together they would talk about it and search out the past. 'I have some lovely memories of Fred. I'd turn up at his house twice a month, sometimes three times a month, in the evening when it was quiet and go down the shed. I didn't end up in the pub with him; I used to sit in the shed and listen to him relate stories to me, and I'd relate stories to him.'

Alan McEwen's boiler works was at Cowling, on the

Lancashire–Yorkshire border between Colne and Keighley. Here he repaired and restored boilers for traction engines, steam locomotives and stationary engines from mills and factories. Fred would be a frequent visitor, particularly in the summer. 'He'd come up here with his number two wife, Sue, and Jack and Roger,' Alan recalls. 'The two little boys would join forces with my lad and my lass, Sean and Alice. And they'd stay playing on the farmstead here, just around the engineering works. But Fred would always like to venture forth. He'd like to find out what was in Keighley, for example, so prior to Fred's turning up I would try to put a tiny itinerary together of places to visit. One was a massive indoor waterwheel over at Mothersdale, which is only a cock's stride away over the hilltops, and Fred was absolutely enthralled with it. The wheel is a mammoth object, built about 1850, and it's incarcerated in its own wheelhouse. Within five minutes of getting through the door, Fred could not only tell me how the wheel had been constructed from three different materials – cast iron, wrought iron and timber – he also informed me that the bearings were made of a very hard wood called lignum vitae. "If you go to a bowling green," he said, "you'll see old guys playing bowls. The bowls are made of lignum vitae." He knew a lot about timber because of his background as a joiner. After that we walked round the mill. I'd been visiting it for a long time but Fred was able to explain to me aspects of the mill that I wasn't aware of, like the hydraulic engineering.

'Another day I had mentioned to him there were a number of interesting chimneys down in the town of Keighley. So we jumped in my truck and went down to this particular chimney and within minutes he was hanging rather precariously from the edge of some stonework, with about a twenty-foot drop into the river Worth below, which was in flood at the time. That was typical Fred, just to get a closer look at the chimney across the beck.'

By this time Fred was getting less and less steeplejacking work. There were very few left to repair and there weren't even many to fell. 'They don't come that often these days,' said Fred. 'Every felling I do now is a big occasion. It gets quite a buzz going. In fact both my wives used to say, "There's no living with him if he's not knocked a chimney down for a bit." It's a thing a lot of people would love to do. It's legalized vandalism really. It's a dangerous business as well but I'm always 99 per cent sure it'll go the right road. As far as the steeplejacking goes I've survived forty years climbing up two-hundred-foot chimneys and I can honestly say I've only had maybe a couple of dodgy moments where I've thought, bloody hell, I shouldn't have done that, I could have ended up dead. You've always got that in your mind when you're doing this sort of work. You don't get complacent, you know. Them men who do, end up dead. When I were a little lad there were literally hundreds and hundreds of chimneys. There must have been at least a thousand in Manchester, Rochdale and Bolton, maybe even more. Most of them have gone now, which is sad.

From the top of one slag heap in Bolton, when I were a little boy you could count two hundred chimneys and now I think you can count what's left on one hand.'

By 1996 there had been a significant change in Fred's life. For around fifteen years, ever since that first phone call from the BBC, there had been regular calls from Jean Thompson to arrange some more filming with Fred. The filming had never provided enough money to supplant his regular job, but along with the public appearances and the odd commercial that his fame brought him, it had been a useful extra source of income. But television moves on and even the most successful series don't go on for ever, so the filming began to dry up. Over twenty films had been made about Fred's life, adventures and marital ups and downs but what more could be done that hadn't been filmed already? He was nearing sixty and his steeplejacking and chimney-felling career, which had provided all the excitement for the films, was drawing to a close.

Fred was now spending more and more time in his garden working on his engines. Neil Carney remembers those days. 'During the five years I worked with Fred I would say that we worked on what I regarded as chargeable work, chimneys and steeples and general steeplejacking, about half the time. The rest of the time we were, to use Fred's own phrase, playing in the garden. "We're playing in the garden today. It's too windy," or "bad weather", or, "We haven't got the right stuff for that job, so we'll leave that till another day." We would just press on with doing bits for his tractor. There was always something to tinker around with in the garden and I spent some happy times there.'

But the tractor wasn't just in the garden. Parts of it had started to invade the house, which was making life difficult for Sue. Even the room designated for a much-needed new kitchen had been taken over by the wheels. According to Fred, the grand plan was to move the wheels out of the kitchen but there was no sign of it happening. 'I did the riveting in there,' he said, 'and when I had them shot-blasted with paint I rolled them in there because it wasn't even part of the house and I put the rubber tyres on in there and then I painted them in there. But then they were all in the way of our new fitted kitchen, so Sue wasn't happy about it.'

And there was an even bigger problem. Working on the tractor didn't bring any money in and Sue was struggling to bring up two young boys on Fred's fairly meagre earnings. Living with Fred wasn't easy and Sue talked about her frustrations in one of Don Haworth's later films. 'A few years ago, way before I was married, I would have expected my husband to do a lot more than Fred does in the house. But I never envisaged the sort of lifestyle that I'm leading now. The general consensus about sharing things in the home I think doesn't really hold for much when a person like Fred, who is doing a very hard job, comes home cold and wet after being out grafting hard all day. Sometimes he'll have busted his thumb because he's hit it with a hammer; you couldn't really expect him then to come in and start doing things in the house. I just don't think it's fair, though. I think you've got to strike a balance, and I think ours is about 90/10. Me for 90 per cent of the housework and Fred for 10.'

Colin Shaw, Fred's old friend from primary school, recalls what it was like. 'I remember being there when Sue was there and Fred was out in the shed making a weathervane for a steeple. I was in the house talking to Sue and she said, "I'll just go out and see how Fred is because his dinner's ready." An hour and a half later, he still hadn't come in and the meal was ruined. He was very selfish in that respect. He just wanted to finish whatever he was doing.'

All this, added to the financial pressures, put a huge strain on their relationship. One of the things that Sue found particularly difficult was the number of days when Fred had got a steeplejacking job to finish, but instead chose to stay in the garden to work on his engine. The marriage couldn't last and when she met somebody else she moved out, taking Roger with her and initially leaving Jack with Fred.

Fred said afterwards, 'She didn't hate me, she didn't depart like some do saying, "I hate you, I'm going to have half of your money" and all that. She went away from here saying, "I don't want your antiques, I don't want your house, I don't want any money; and I'll not leave you in the shit with all the books and typing. I'll get somebody in to do all that for you." But what she did do was take my lads away from me and I can't forgive her for that.'

Fred was on his own again and he found it just as hard as the first time. His house was on the edge of Bolton's red light district and he told me about one rainy night when he saw one of the girls trying to shelter just outside his house. He'd seen her around a few times before and knew her to say hello to, so he asked her if she wanted to come in for a cup of tea. They talked for an hour or more and as the girl left she said to Fred, 'You know where I am if you ever need my services. And it won't cost you anything.'

But that was not what Fred wanted. He was looking for a new lady to share his life, and he knew what sort of woman he wanted. She'd have to be interested in antiques and steam engines and beautiful things, he said. 'As well as liking the world of Queen Victoria she'd have to be able to answer the telephone and be a bit diplomatic with people who ring me up and want me to do all sorts of unbelievable things that I don't want to do. And she'd have enough sense to suss the nutcases out from the genuine ones and get it all written down on a bit of paper. She'd have to dress well because I like nice dresses, preferably the ones that fit well, and black nylons and high-heel shoes and all of that sort of thing because I think black nylons and high-heel shoes enhance a lady's legs, make them look nice. She'd also have to be a good cook and keep the house nice and clean and tidy.'

Fred was not the modern man and some of his views, especially when it came to women, were old-fashioned and unenlightened. The woman's work of keeping the house clean was what Fred had grown up with; his attitudes and values were shaped by his background and environment and by his Victorian sensibilities, which made him something of a male chauvinist. He always claimed that he had been born out of his time. He reckoned he should have been born in the Victorian Age. 'This equal rights thing,' he would often say, 'isn't something I agree with. What they want now is half of everything. Difficult, isn't it?' Fred was always one of the lads and he never wanted other men to see that he wasn't master of his own house or of his own time. If Fred had got work to do on his engine or he wanted to spend time drinking with his mates, that was his prerogative, he felt. Cooking and housework or even taking his wife on a proper holiday never came into it. Fred could be difficult to live with and for each of his three wives these attitudes led to problems and confrontations.

For Fred, finding the perfect lady would have been a tall order even in Victorian Britain, but as we approached the millennium how many women were there who would meet his requirements? And, in spite of his fame, how difficult was it going to be to find the perfect partner for the third time? After all, recalled Colin Shaw, 'Fred had never really been one for the ladies. In all the years that I knew Fred I never saw him at the Palais or the tennis club or any of the other places we used to go to meet the girls.

At school I never associated Fred with any girls and even in later life I didn't. But, thinking about it, he always needed somebody to look after him. It was somebody to make his meals; it's as simple as that. And on his own he'd never make any money; he needed somebody all the time pushing him one way or another.'

The winter of 1996–7 was, as Fred recalled many times later, a very difficult time for him; no wife, no money and no prospects. But the situation would not last for long. He was about to find a new partner who would not only make his meals but give him that little push that Colin Shaw said he'd always needed, and he was set for a return to the television screen that would bring him even greater fame and more money than he'd ever thought possible.

Part Two
The Working Man's Hero

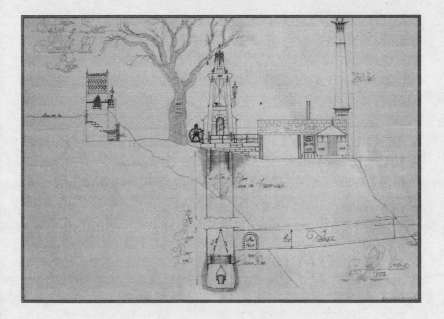

7

A Fresh Start

My first meeting with Fred was on a wild and windy night in January 1997. It came as a result of a phone call from Frank Fallows, who knew Fred from the time he'd worked for the distribution company that handled a couple of videos about him. He'd spent some time with Fred promoting the videos and after he left the company they kept in touch. By the time of Frank's call to me Fred was at a very low ebb. He was desperately sad that his boys had been taken away from him and financially, although Sue hadn't asked for any money or any share of the house when she left, Fred was in dire straits. He told me later that at the time of our first meeting he'd got £240 in the bank. Fred rang Frank to ask if he knew anybody who might be able to do something to resurrect his television career.

Frank phoned me immediately. After the call I talked to Steve Parry, the editor I was working with. People who work in television are not usually star-struck but Steve's reaction was 'Honestly, Fred Dibnah? He's great! I never missed any of his programmes. You've got to go and see him and try to get him back on.' (Little were either of us to know that Steve would go on to be the unsung hero of more than thirty of the films I made with Fred, cutting out hundreds of 'yeah, yeahs' and 'you knows' from each of them, putting together as many as three different takes of Fred to get the best possible piece to camera and creating memorable sequences like Fred's action replay of Walmsley's rolling mill in Bolton that had been pre-served at Ironbridge. Throughout all the years of editing those films, Steve remained one of Fred's biggest fans and later it was Steve who nominated Fred for his MBE.) On that January afternoon I took Steve's advice and made arrangements to go to see Fred.

When we got to the house it was immediately clear, even in the dark, that it was a bit out of the ordinary. It looked like the gatehouse to a big Victorian mansion, all very ornate, with lots of nice stonework and fancy bits of stone and ironwork on the roof. At the front, where we parked, it appeared to be single-storey. We didn't go to the front door because that was the upper floor where the bedrooms were, so Frank guided me down some steps at the side to the back door which led out into the garden. In the dark my first impression of the garden, where I was to do so much filming, was that it looked like a junkyard. But, when Fred opened the back door, the interior of the house had a very different impact. It was like stepping back into the Victorian Age. Fred led us through what looked like the original kitchen, which was dominated by a magnificent black-leaded range – the sort of thing I'd last seen when I visited my grandmother in the 1950s. The room had a decorative cornice and gas lamps on the wall, lots of shiny brass and old framed pictures of traction engines, chimneys and industrial scenes. Fred was clearly very proud of his house and the beautiful things in it. The parlour was very similar in its décor, with two leather chairs and a green leather chesterfield grouped around a large, ornate fireplace. A fire was blazing in the hearth. The only concession to modernity was a big television in the corner with a pile of what looked like steam videos on the floor beside it.

Sitting in one of the armchairs by the fire was a boy of about nine. It was Jack. Roger was over in the Isle of Man with his mother, who had moved there to live with her new partner, but Jack was back in Bolton with his dad. He loved the garden and the engines in it. He'd grown up with them and played with them from the time he could walk and he was already showing the engineering skills that were to blossom as he got older. Fred sat down in the chair opposite Jack and they looked like peas in a pod. Jack was a mini, slimmed-down clone of Fred, complete with a cap just like his dad's. It was an image of a father and son from a different age and I couldn't help thinking of Jack as a little Victorian chimney boy, ready to get up the chimney when the fire had died down and his dad had got his brushes out.

Instead he listened quietly as Fred began his tale of woe.

I'd been warned by Frank to be prepared for a long session. Fred could talk, he said, and once Fred had started that night he talked for over three hours non-stop. He was one of those people who always needed somebody to talk to and he was deeply lonely. Even though Jack was there that night, he was distraught because his two lads, his pride and joy, had been taken away. I was a complete stranger but Fred poured forth his soul that night, as I was to see him do with many other strangers in the years to come. Here was a man who clearly wore his heart on his sleeve.

He began by telling us how difficult the steeplejacking and chimney-felling business had become. At a time when there were fewer and fewer chimneys to work on, there was more competition, with lots of men from Manchester prepared to undercut his rates to get the few jobs going. Then there was the television, or the lack of it. The filming days had gradually tailed off and over the last few years he'd not had any phone calls from Jean Thompson. It seemed that the BBC didn't want any more Fred Dibnah and anyway, he said, he'd never made much money from it. Then he talked about the breakdown of his second marriage and became very upset.

Suddenly he cheered up a bit and said there was a new lady on the scene. This tall blonde, as he described her, had turned up at his house recently and shown an interest in his engines and workshop. Sheila Grundy had arrived one day with her mum and dad and her little boy, Nathan.

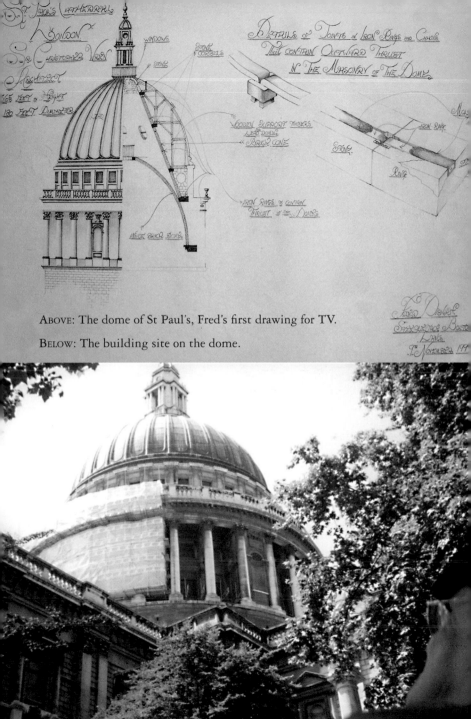

ABOVE: The dome of St Paul's, Fred's first drawing for TV.

BELOW: The building site on the dome.

The summit of Fred's steeplejacking ambitions.

Always time to stop for a photo and autographs.

'It's a bit stinky down there' – the latrines at Housesteads Roman Fort on Hadrian's Wall.

Blackpool Tower and
Fred's representation of
how it was built.

BLACKPOOL TOWER
MAXIM AND FROUDE
MANCHESTER
5/8 OF HIGHT

GUY ROPES

PULLEY BLOCKS

WOODS FOR PLY

HAND WINCH

STEAM HOISTING ENGINE

FRED DIBNAH ©
STEEPLEJACKS
BOLTON

Fred with his friend
James Hervey-Bathurst
of Eastnor Castle and
his engine *Atlas*.

A bit of steeplejacking
while filming at
Glamis Castle.

ABOVE: Fred's garden, filled with pithead gear, winding engine and chimney.

RIGHT: Fred's shed, the place where he was happiest.

OPPOSITE PAGE, CLOCKWISE FROM RIGHT: Fred always welcomed visitors without charging them, but anyone who wanted to contribute could put something in the 'Guinness fund' collection box; with Roger and Jack; a single container of pansies was the only woman's touch; back-street mechanicking.

ABOVE: A great day out at the Strumpshaw Steam Collection.

BELOW: Ready to drive Tom Brogden's replica of the Hanson Steam Carriage.

CLOCKWISE FROM TOP: Preparing for a take with David Hall; hands-on in the chainmaker's shop at the Black Country Living Museum and at Finch Foundry; on the Great Western at Didcot Railway Centre.

From top: Levant Mine and Engine, Cornwall; Hook Norton steam-powered brewery, Oxfordshire; ready to drive *Fenchurch* on the Bluebell Railway, East Sussex.

They were all looking down over the railings into the garden, so Fred had invited them, as he did with everybody, to come and have a closer look. Sheila was a real good looker, he said. He thought she'd been a former Miss Blackpool and on the stage as well as an assistant to some well-known magicians and hypnotists. But what was more important for Fred was that she seemed to be really interested in everything in the garden.

Sheila recalled that first meeting. 'We went down into the garden and stood there for about two hours listening to Fred, who was so enthusiastic about everything. He was telling me all about his engine and about where things had come from. I could see this great passion he had straight away. It was a freezing cold day and I recall that we signed the visitors' book like everybody else did and went away and thought that was the end of it. A few days later, though, I was still thinking about him, because he's got a magical effect on you when you first meet him. I sent him a small note, but I don't even think I put my address on it. So Fred looked at his visitors' book and he contacted my mum by phone. He said, "Ay, that young lass of yours like, she lives in Bolton sort of style, does she?" He rang me up a few days later and then sent me a letter asking me to come down again if I wanted and to bring my son with me. I had another look round the garden and we became friends because we'd got a lot in common at that time. Fred had just gone through his second divorce, and I'd gone through a bad relationship, so we had plenty to talk about on a personal level, and he was a really interesting man.'

By the time Frank and I went round to visit him, Fred had seen Sheila a couple more times. He was clearly flattered by her interest. She was good-looking and she was twenty years younger than he was so what was it she saw in him, he wondered. It certainly wasn't his money because it was obvious he didn't have any. She seemed to be genuinely interested in steam and steeplejacking and all the bits of industrial archaeology that Fred was so passionate about, and she liked things that were old. It was early days yet but Fred, ever the optimist, was wondering whether this time he'd found the right woman.

Fred was much more cheerful by now and he began to launch into some of the steam and steeplejacking stories that I was to become so familiar with over the next few years. He talked about Bolton and the time he was growing up there. I'd spent my earliest years in the Bradford district of Manchester and I told him about my memories of Bradford Pit, which was just down the road, and of walking past Johnson's wireworks with my dad on our way to Phillips Park. As we talked it was evident that the industrial north-west was where his heart was, especially the Bolton of the 1950s and 1960s. He spoke about the glory days of King Cotton and how it had played such a big part in shaping his world and he told me stories about how, as a little boy, he'd climb up and look into engine houses at great big shiny steam engines. Perhaps if he could have turned back the clock he would have lived in the 1920s or 1930s and been an engine driver in a spinning mill or something like that.

His passion for everything connected with Britain's industrial past shone through. From the huge stationary steam engines that powered the pumps and pithead gear of nineteenth-century collieries to the clattering wooden-frame machines of the textile mills, he'd got a story to tell about the people who made them and operated them and he had a vast fund of knowledge about how they were built, how they worked and how they were used. He wanted to get back on television to let people see some of these things before they all disappeared.

So far, in all his BBC programmes, Fred had been the subject of documentaries. I told him I thought it would be difficult to get more films about him and his life commissioned. He'd had a good run but, if we were to get him back on television, it would need to be with something different that would build on his passion and his popularity. We started to talk about a series that would give him an active role as a presenter and draw on the knowledge and enthusiasm he had for all matters relating to industrial archaeology and the age of steam, and on his contacts with people and organizations who were involved in restoration. It wasn't something he had ever done before, but he was keen to have a go.

I went away from that first meeting and started to think about industrial history on television. Although there were a lot of history programmes, industrial history was at that time a relatively neglected area. But there was a real need to do something. Britain's industrial landscape had changed dramatically over the previous thirty years. Whole industries had just about disappeared and the communities that had grown up around them had been forced to adapt to new types of employment or, in many cases, to no work at all. By the mid-1990s industry had become a bit of a dirty word. John McDougall, a past president of the Society of Mechanical Engineers, summed up the climate of the time. 'I think especially through the Thatcher years manufacturing was neglected. It was thought that our economy could be regenerated using the service sector only and what we forgot was that somebody has to make that original one pound before the service sector can gear it up to many more pounds.'

The idea that emerged for that first series was for an enthusiast like Fred to undertake a tour of Britain to uncover its industrial heritage and meet some of the often eccentric characters who devote their lives to finding, preserving and restoring steam engines, mill workings, pumps and pit winding gear. He would take us to the places where the Industrial Revolution had its birth and the parts of the country where significant developments occurred. His trail would include historic industrial sites, museums, industrial heritage centres and the workshops of preservation societies who were excavating sites and rebuilding the machines housed in them.

Despite Fred's lack of presenting experience, the more I worked on the project the more I became convinced that he was the man for the job. His approach would be very practical. He would be there as the enthusiast who didn't mind getting his hands dirty as he fired up the steam engines, drove the traction engines and steam locomotives, got down to the coal face and took machines apart to show us how they worked. Fred would be an expert with a mission to tell the story of the Industrial Revolution and to highlight the work that was being done to preserve our industrial heritage; an enthusiast with a passionate interest in the places he would be visiting and the activities he would be taking part in rather than just a reporter. Above all, Fred was an entertainer and whatever he was talking about he'd do it in an entertaining way.

As I did some research on possible locations, things began to look up a bit for Fred. He was still doing some after-dinner speaking and he also did a commercial for Kellogg's Corn Flakes. The scenario was a house in the middle of a cornfield and it involved a man working on the roof. Fred, of course, was the man on the roof. It was being shot in South Africa and Fred flew out with the cast and crew. He wasn't a great traveller, in fact he'd not been out of the country since his National Service days, and neither had he worked with actors and models before. But he enjoyed himself immensely up on that roof in the South African sunshine. It was a bit like a holiday that he was getting paid for. He also found, somewhat to his surprise, that he got on very well with one or two of the actors. They weren't arty types at all, as he'd thought they might be, but ordinary blokes you could go for a drink with and have a laugh.

By the middle of the year, Fred was seeing a lot more of the 'tall blonde' and Sheila was finding him more and more interesting. 'Fred's not exactly like a Tom Cruise character,' she said later, 'but I was just sucked in by his personality and his enthusiasm. He was different from anybody else I'd ever met and I just wanted to get to know him a lot better. He invited me along to a chimney-felling operation and to a steam rally as well. The organizers put us up in a tent, with a pair of bus seats for beds. It didn't make the best impression.' Fred certainly hadn't lost his touch when it came to charming a lady.

Later that year I sent a package of ideas to the BBC Factual Entertainment Department. One of them was for a popular, entertaining approach to industrial history and I put Fred forward as a possible presenter. Within a couple of weeks they'd come back and said they would like to go ahead with a series of six half-hour programmes for BBC2. They wanted Fred to present and would commission it subject to his being available and enthusiastic for the series and its proposed content. It was a format that would give Fred's fans the opportunity to see him on their screens again. This time he'd be there in a different guise – as an expert with a passion and a purpose. They just needed something in writing as soon as possible to confirm, subject to contract, Fred's involvement with the series. No problem. It was just what Fred wanted and I couldn't wait to ring him with the good news.

When I rang the number Fred had given me, it wasn't

Fred who answered. By this time he'd got a new assistant, Simon Warner. Simon had been working with Fred for some time but since Sue had left he'd been helping with the general running of the business as well as the actual steeplejacking and felling. Simon informed me that Fred wasn't available and that if it was an enquiry about television work he'd now got an agent to look after these things. I would have to ring a lady in Mansfield to discuss the proposal with her. Claire Brackenbury had, it seemed, been brought in by Sue before she left, because once she'd gone Sue knew Fred would need somebody to help him manage his affairs.

I rang Ms Brackenbury straight away and was surprised to be told that Fred wouldn't be able to do this series. He was, she informed me, already doing a new television series. There was no broadcaster involved but she had asked Nick Wilding, a producer she knew, to do some filming with Fred.

Eventually, after a number of phone calls, Ms Brackenbury agreed to a meeting at which I could outline the proposals for the BBC series and, as long as Nick agreed that it didn't cut across anything that he was doing, it could go ahead. The meeting was at Claire's house and when I was ushered into the dining room I was confronted by a roomful of people crammed round the table. I can't remember how many were there, but the group included Claire's husband, Nick Wilding, Simon Warner, Sheila and Fred. I knew that Fred's most pressing need was to earn some money and that there wouldn't be very much for him in the purely speculative filming that was being done, so I immediately quoted the fee I had put in the budget for him. I'd pitched it high and, as soon as I mentioned the figure, the mood of the meeting changed and everybody seemed to want the BBC series to go ahead. Fred was sitting by my side, a look of shock bordering on disbelief on his face. Nick, to his eternal credit, soon agreed that there was no clash with the filming he was doing. It was great that Fred was going to be presenting pro-grammes, he said. It really could be a new start for him.

As Fred drove back to Bolton with Sheila that after-noon, he couldn't believe the change in his fortunes over the past year. He'd got a nice-looking new lady by his side who was interested in all the things he was interested in and now he'd not just got back on television but he'd been offered far more money for one job than he'd ever earned for any work he'd done in his life before. Fred had a smile on his face again.

We joked about that meeting many times after that and about how close he'd come to losing not just that series but everything that followed on from it. But I only found out later that Fred had been in grave danger of losing a lot more than a mere television series. In September 1997, a couple of weeks before my phone call, Fred had under-taken a major chimney-felling. It was a 450-foot concrete monster on Canvey Island in Essex – far higher than any-thing Fred had ever done before. It had been built in the late 1960s for a power station that had now closed down and it needed to be cleared to make way for a new

Safeway supermarket. It could, and probably should, have been a job for the dynamite boys, but the contractors wanted to put on a bit of a show so good old Fred had been called in.

Fred had never attempted a concrete chimney before and it was a very different proposition to the brick chimneys he was used to. He went down to look at the job and decided that, apart from needing to take care with the steel reinforcing rods inside the concrete, he'd have no trouble with it. Fred travelled to Canvey Island with Simon Warner and his long-standing felling assistant, Mick Berry, at the beginning of September and they started their preparations. As far as Fred was concerned it was just a normal job requiring the usual cut-and-prop technique that he'd used so many times before. The only difference was that they would have to drill and cut the steel rods all the way round the base.

Everything was going to plan. The rods were cut and the concrete on the side of the fall was cut out and propped. As they worked, a scaffolding grandstand was erected for the invited spectators and the day before the drop there was the usual buzz of excitement and anticipation. Fred and his boys were finishing off their preparations at the base of the chimney and checking all the props when Fred noticed tiny specks of concrete drizzling down on them. They carried on drilling a bit longer but the tiny specks started to get bigger and within seconds great lumps of the stuff were raining down. Fred yelled, 'Run! We're not in control of this bugger.'

Instinctively the three of them got round to the other side of the chimney and raced away in the opposite direction to the planned fall. But this stack was volatile and there was no knowing whether it would go in the direction it was supposed to. And they'd no experience of anything like this. Mick and Simon got a head start on Fred and then disaster – Fred tripped on a wire and went flying. He was only about fifty feet from the base of the monster, which was ready to go at any second. Fred said he would never forget that moment as long as he lived.

'It was like watching one of those films where everything goes into slow motion. I tried to get up and get further away, but I seemed to be frozen to the spot, looking up at about 2,500 tons of concrete and steel ready to come crashing down. Then, after what seemed like an eternity, it started to go in the right direction, away from where I was.' It had been one of the closest shaves Fred had ever had. After a few pints he soon recovered and even claimed he should have had a bonus for getting it down a day early. But with a close call like that his new television presenting career very nearly didn't get off the ground.

The summer of 1998 was the time for filming the series, so the research started in the spring. The first thing was to meet up with Fred to find out what he wanted to do and whether there were any particular places he wanted to visit. He didn't have any strong feelings because until that time he'd not travelled far from Bolton very much and he wasn't aware of most of the industrial heritage sites around the country. He'd been to a few of the better-known ones like Ironbridge and he was keen to see George Cushing's traction engine collection at Thursford in Norfolk. He also mentioned a few local places like Wet Earth Colliery, Astley Green Mining Museum and the Northern Mill Engine Society. But apart from that he was happy for us to find all the locations for the series and sort out the content.

By this time Sheila had moved into Fred's house with her son, Nathan, and she soon found out that Fred's expectations of a modern woman were quite different from anybody else's she had ever come across. 'He got quite excited once because we were watching a programme on TV and it showed women down the pit. I was horrified to see these young lasses of about eighteen, with big belts and chains around them, shackled to a pit tub with a hundredweight of coal in it. But Fred thought it was great and said, "There you are, I should have married somebody like that!" ' Just the right sort of attitude for the presenter of a new industrial history series. At least it would be a bit more colourful than the usual more academic approach of television historians.

Before we started filming Fred was still getting some steeplejacking work which Simon Warner was doing with him. He laddered a church in Preston to have a look at the extent of the work to be done on it and he was still being offered a few chimneys. Sheila even climbed up a chimney with Fred one day. 'I'll never forget that one occasion,' she recalled. 'It was Easter Monday 1998, and the oily gauntlet had been thrown down because I'd said to Fred in the company of someone else, "Well, anybody can climb a chimney." That was it. Somebody piped up and said, "Well, you couldn't go up one." Of course I could, I said. Then I had to prove I was as good as my word, so I climbed up to the top of the chimney. What sticks most in my mind about it is that when I got near the top and I was holding on to the ladder with grim determination, Fred, who was just down below me, said, "Let go a bit, that's the grip of death you've got." But of course it was. What else would you expect when you're nearly two hundred feet up in the sky? I did manage to make it to the top, but what a frightening experience it was.'

The chimney was frightening for Sheila, but Fred, as I learned later, was starting to get a bit frightened about something else. It was the series. Although he'd had a camera around him on many occasions and was used to having a film crew track his every move he'd never had to stand in front of a camera as a presenter, follow directions and remember what to say. It was very different to anything he had done before. I too was beginning to wonder what I'd taken on. I knew Fred was a good talker and that I'd be able to give him some information or get him to read something, then put him in front of a camera, wind him up and let him talk. I also knew he'd be able to ad lib very well and tell us how things worked, and that with skilful editing I could cut round what he was saying and make his explanations easier to understand. But the sort of programmes we were talking about would have to be quite structured, and structured programmes need a lot of scripting. It was this that Fred was worried about.

Before we started to do any filming there was a book to be written. BBC Books had already had a success with Don Haworth's book *The Fred Dibnah Story* and they had commissioned a book to accompany the new series. The idea was that I would write a popular history of the Industrial Revolution which would be interspersed with Fred's anecdotes. It would go out in joint names with me doing the writing and Fred doing any promotion and personal appearances related to it. I went to Fred's house with a tape recorder to interview him and get some of his stories for the book. We sat on the chesterfield in his parlour and Fred started to talk about his earliest memories of industry in Bolton and of the railway sheds at the bottom of his road. He painted wonderful, evocative pictures and, as I sat there listening, I could see the sheds he was describing and picture the locomotive flashing past the end of his street in the middle of the night. I heard for the first time all his stories about what he got up to with his mate Alan Heap and their mad inventions like the diver's helmet they made out of a Smith's crisp tin.

Fred really could tell a good story. But then he got on to talking about the work he'd done on his traction engine and the restoration of the engine at Glynllifon. It was a warm afternoon in early summer and as Fred went into minute detail I began to feel drowsy. I had absolutely no idea what he was talking about and as he rambled on I started to nod off. Fred was so preoccupied with the technicalities that he didn't appear to notice, or at least he was too polite to show that he had. Eventually I left with three tapes full, thinking that my falling asleep had hardly been the most auspicious start to our working relationship.

I took the tapes back to my office and gave them to an audio typist to transcribe. When I got the transcriptions back, the first thing I noticed was that there were hardly any full stops in them. Fred rarely paused for breath so one sentence just ran into another. This was fine when he was giving his beautiful descriptions of the engine sheds or of the Bolton skyline when he was a little lad, but it became impossible to decipher once he started talking about the engines. It was like a foreign language. After spending hours trying to work out one little section I began to see what was going on. Fred had a habit of going off at a tangent. He'd start off talking about engine A, let's say his traction engine. He'd get into a lot of technical detail about a particular part of the engine and then without a pause for breath he'd start talking about that same part on engine B, let's say the Glynllifon engine. Without giving any indication that he was talking about another engine he'd then move on to the detail of a third engine or go back to the original. At no point in the transcript was it possible to detect when he'd moved from one engine to another.

Fred was clearly obsessed by detail. Just as he had

wanted to give that detailed description of the way he laddered a chimney in the first Don Haworth film, he still felt the most important thing was to explain the finer workings of the machinery we would be looking at. There was no cog so small, no crankshaft so insignificant, that it didn't deserve to be elevated by Fred to the status of an oily jewel. But the level of detail he wanted to go into was of only limited appeal. It was the sort of thing the real enthusiasts would want, but it was too much for a popular book and it certainly didn't have a place in a prime-time television series. Fred had clearly got all the passion and enthusiasm that was required but it was going to need a lot of directing.

Before we started any filming I talked to Fred about the way we would go about things. The programmes needed to be scripted but he was very uneasy about having to learn scripts, so I suggested that we wouldn't have scripts as such. Instead I would give him a set of notes for each location. These would contain the key points he needed to make and they would all be broken down into little segments. He could look at each short piece and say it in his own way as long as he got the sense of it right. In addition we would film him talking to people at the places we went to, and if we needed to know how something worked I would ask him to ad lib a bit and explain the mechanics to me in as short and simple a way as possible. I said I was pretty dim when it came to anything mechanical, so, if I could understand it, just about anybody could.

When he knew that he wasn't going to have to learn great long scripts off by heart Fred felt a lot easier about the filming. The script notes we did for him became the basis for all the programmes we made over the next seven years and we developed a way of working that involved Fred doing his pieces to camera one or two sentences at a time. Fred became very adept at looking at the notes and delivering each of the bits of information in his own inimitable style. He'd then do several much longer pieces, telling a story or describing how something worked, and the elements would all be brought together in the edit, along with the interviews he'd done and the scripted voiceover commentary which he would record when the films had been edited. Fred Dibnah, steeplejack and steam enthusiast, was ready to become Fred Dibnah, television presenter.

8

Industrial Age

Filming *Fred Dibnah's Industrial Age* began in July 1998 and for the first few weeks it was very much a case of trial and error. When Fred received his schedule for the first week's filming he rang me to ask about travel and accommodation arrangements. In the proposal for the series Fred was going to be seen driving around the country to all these industrial heritage sites in his Land-Rover. Naturally he thought he would be doing this for real, but his Land-Rover was very slow and very heavy on diesel. It couldn't be used for a filming schedule. Instead we'd do all the driving shots around Bolton some time towards the end of the shoot but for travelling to the locations we would pick him up from home in one of our production cars. As far as accommodation was concerned he wasn't to worry about it; it was all provided for him and had already been booked. Fred was amazed. Not only was he being paid for this trip round the country to all the places he was really interested in, but all his travel and food and accommodation were going to be provided as well. Whenever he'd worked away from home before he'd had to find his own way there and get his own digs. This really was a change.

The first day's filming, however, was on home ground. Fred had always had a great interest in mining and a lot of his friends were retired miners, so it was appropriate that our first two locations were coal-mining sites not far from home. Wet Earth Colliery is in the Irwell Valley between Bolton and Manchester. It's important because of the work that the eighteenth-century engineer James Brindley did there. The colliery had been closed for many years but a group of enthusiasts known as the Wet Earth Colliery Exploration Group had got together to excavate some of the workings. This was real industrial archaeology. They met every Saturday morning and it was their activities that we were going to film.

That first morning's filming gave a good indication of the sort of thing we had in store. The exploration group had found that Brindley's scheme had relied on water power to drive a wheel, which in turn operated pumps to drain the mine. One of the things we were filming was a sump that Brindley had dug under the river Irwell to channel water to the great wheel that provided the power for the mine. It was an interesting piece of eighteenth-century mining engineering and Fred was keen to have a look at it. Alan Davies from the Lancashire Mining Museum and his team of volunteers had the gas detectors that we needed to go into the underground tunnel that led to the sump, so everything was fine from a Health and Safety point of view. Or so we thought.

Once we got into the tunnel it was pitch dark, illuminated only by the miners' lamps. The muddy ground under our feet sloped away slightly from the tunnel entrance and ahead of us wc could hear the sound of running water. It was the sump. No railings there, no protection; just a slippery slope and a twenty-foot drop into the water with no way of getting out of it. One slip from anybody and they'd be in trouble. Fred walked to the edge of the abyss unconcernedly and stood there chatting to Alan, wondering why Brindley had needed to construct the sump in the first place. In order to get a shot of what they were talking about, cameraman Mark Hamilton had to get even closer to the edge to point his lens down the hole. Fortunately Alan had a rope with him and, with the rope tied securely round his waist and several people holding on to the other end, Mark got the shot that was needed. It was a good indication of what was to come with Fred.

That afternoon we visited another local mining site that Fred knew well, Astley Green Colliery Museum. Fred was keen to have a look at the steam winding engine there, which is the largest in Europe. The pit closed in 1970 but a lot of the surface workings were left standing and a group of enthusiasts were busy trying to restore the winder and some of the other pithead machinery. When we arrived at the site we started to unload the kit for a couple of pieces to camera from Fred. After checking the script and sorting out where I was going to do the shots, I looked round for Fred to run through it with him. He'd disappeared. Astley Green is dominated by a one-hundred-foot-high pithead winding gear. It's a magnificent sight but it's clearly seen better days. Years of lack of maintenance and exposure to the weather had left it looking quite derelict and dangerous. Not the sort of thing you would want to climb up in a hurry – not unless you were Fred, that is.

We'd not been on the site for more than a couple of minutes and Fred was nearly at the top of the winding gear. How he'd got there I don't know. There was no ladder for the first fifteen feet or so, presumably to stop the foolhardy from trying to climb it. But that wasn't going to stop Fred. Here he was, a sixty-year-old, racing up to the top of this rusting metal tower like a kid. It didn't look very safe and as the director of the shoot I was responsible for Health and Safety. But Fred clearly wasn't going to let Health and Safety get in the way of doing what he wanted to do. Should I get a shot of him up there? It would certainly look good. Better not, I thought. If I do a shot, it will look as though I've directed him to go up there, and then if he has an accident I'll be in trouble. By this time Fred had reached the top and was walking round on a small platform. 'Are you coming up?' he yelled down to me. I made my excuses and tried to look busy around the camera. When he came down eventually to start the shooting he was beaming. 'I really enjoyed that,' he said. 'Did you get any pictures of me?'

Later that afternoon we set off for the long journey to the West Country, where we were to do our first week's filming. The next day we were at Coldharbour Mill in the beautiful Culm Valley near Cullompton in Devon. The mill has two floors of working Victorian machinery, a waterwheel that was being restored by volunteers, a couple of steam engines and a Lancashire boiler. Once Fred had got a couple of pieces to camera out of the way he was in his element, being filmed meeting some of the volunteers. The highlight for him was climbing into the Lancashire boiler with the film crew to have a look at what it was like on the inside. That evening, as I sat having a pint with him at the bar of the hotel where we were staying, he said, 'I'm enjoying this. It's a bit like a holiday really.' And what better place for a holiday than the West Country?

We spent the next few days filming around the old tin mines and engine houses that give the Cornish landscape so much of its distinctive character. Fred loved it and he loved the nights in the pub after. It was the time of the 1998 World Cup. There were a few of us on the production team who weren't going to miss any of the matches, so we tried to get Fred interested in football. The morning after one particularly heavy session in the pub we were driving from Penzance to Geevor Tin Mine on the other side of the peninsula. Fred was sitting next to me and as we drove down the Cornish lanes I said to him, 'I hope we don't get stopped for anything. I'm probably still over the limit from last night.' It was quite early and there was a thick sea mist. We missed a signpost and took a wrong turning so I stopped just round the corner to check where we were. As I walked back to the corner to look at the sign I heard a car in the distance. It was clearly travelling very fast. It emerged from the mist and screeched round the corner. Brakes squealed but the driver couldn't stop and he ploughed into the back of our parked car. The young man who was driving was full of apologies. He was late for work. Insurance details were hurriedly exchanged and on inspection the damage was not as bad as we had first thought, so we were soon on our way to Geevor and filming Fred doing the underground tour. He showed no sign of the previous night's session. It was the way it would always be. However long the lock-ins the night before, Fred was always first up the next morning, bright and early and ready to appear on camera.

The tin mine was right on the edge of the cliffs near Pendeen. There had been mining on the site for over three hundred years and at its peak it had employed around four hundred men. It had closed in 1990, when mining operations became uneconomic. We'd come here with Fred because it is now the most complete example of a tin-mining and processing site surviving in Cornwall and one of the best mining heritage sites in the whole of Britain, with most of the buildings and machinery still intact and exactly as they were when the mine closed.

Geevor Mine in the mist was an eerie experience. The

vast processing plant where the ore was turned into tin was deserted, but all the machines were still there and the place looked very much as it must have done a few years earlier when the last shift finished. You could almost feel the ghosts of those workers still in there. Our filming in Cornwall in 1998 came right at the end of the once great tin and copper mining industry that had provided employment for many generations of Cornishmen. As we filmed, the last of the tin mines was still operating, but it was to close down within a year. The series we were making wasn't going to save any of these industries, but at least we would provide some record of them and make the public aware of them. It gave Fred a real sense of purpose.

A mile or so along the cliffs from Geevor is the Levant Mine. It was sunk 1,800 feet down and the workings went out for several miles under the Atlantic Ocean. They had long since been closed, but the engine house with its little winding engine still stands on the edge of the cliffs. It's a dramatic sight with the waves pounding on the rocks nearly two hundred feet below. Fred was very impressed – not by the scenery but by something else he saw. 'If you look down into this great chasm,' he said, 'you can see various flights of stone steps coming up the cliff side. In the olden days, before steam winding and wire ropes and cages, the miners had to get down the face of the cliff right near the sea and then enter by an adit that met the main shaft and continue the journey for one thousand and eighty feet on ladders down the shaft. Then they had to go for a mile under the ocean before they actually started work. They must have been some special men.

'Later on they got a steam winder and the miners were carried up and down on the man engine, which was a sort of steam-powered wooden lift. It had a vertical wooden beam with small wooden platforms on it and as it went up and down the mineshaft, the miners hopped off these on to platforms attached to the side of the shaft. But riding on the beam was very dangerous. With everything wet underfoot and trying to do it after a twelve-hour shift, you can imagine what a risk it was; but anything was better than climbing a twelve-hundred-foot ladder to the surface. Then in 1919 there was a disaster. Forty miners plunged down the shaft as the beam that they were riding broke. Thirty-one of them were killed and it took four days to get all the bodies out.' Fred always had a great interest in disasters and he told the story with relish.

What makes this mine so special today is that it contains the first steam engine to be preserved on site. The winding engine had been restored by a group of volunteers from the Trevithick Society, nicknamed the Greasy Gang. When we went to film it with Fred, we were on a tight schedule and straight away we came up against the sort of PR issues and politics that were to get in the way of a lot of our filming with him. First of all there were interviews with the local radio station and photo calls for the local press. That wasn't too much of a problem. Journalists and press photographers in general are very good and will work round the needs of a film crew, so it wasn't too difficult to let them have Fred while we were setting up the shots. The real problem, and we would come across this many times, was the number of people who had been invited along to meet Fred and, unknown to us, to be filmed with him.

The engine house was small, but it was packed with every surviving member of the Greasy Gang. They all wanted to talk to Fred and be filmed with him in front of the engine they had restored and Fred was only too happy to oblige. Technically it was almost impossible to do it. We couldn't cover them all for sound and filming a beam engine in a small engine house is difficult at the best of times, let alone when you've got half a dozen bodies crowded in front of it. But I didn't want to offend any of these elderly men who had devoted so many hours of their lives to restoring the engine and I knew that Fred certainly wouldn't want to, so we tried to work round it. Most of what we shot that afternoon wasn't usable and we returned to Levant a few years later when we were doing *Age of Steam* to cover it properly.

That first lot of filming in Cornwall was very enjoyable for everybody, not least for Fred. During the day we were going to interesting places and in the evening Fred was good company down at the pub, where he was getting a new audience every night for his stories. We quickly developed a good working relationship and many things were learned that would provide the basis for all the filming we were to do together in the future.

Tolgus Tin was an ore-crushing and separating plant we had arranged to film. As we were driving there Fred noticed a small sign for a collection of winding engines and old mining machinery. He wanted to have a look but I said we'd got a schedule to stick to. When we reached the car park at Tolgus Tin, Fred was approached by a youngish man with a photo album as soon as he got out of the car. I could see Fred looking at the photos and talking animatedly. As we were unloading the gear from the car Fred came over and said excitedly, 'This is the man who owns that museum we went past. He's got some lovely engines there and he's restoring them all himself, but he's got no grants for it and he gets no visitors because nobody knows about it. Let's go there and film him.'

Here was a dilemma that was going to come up many times and in many of the locations. The reason we had Fred presenting these programmes was his passion and enthusiasm. Spontaneity was very much part of that, but filming schedules don't allow for that level of spontaneity. Locations have to be researched and recced, arrangements have to be made and permissions obtained for filming at them and you can't just not turn up at a place because at the last minute you've found something else that you want do instead. So I had to say no. We'd do our filming at Tolgus Tin, as arranged. I'd go and have a look at the museum with Fred after that, if it was still open.

Not surprisingly under the circumstances, Fred was a bit disappointed with Tolgus Tin. It was steam that was his real passion after all and he would have preferred to be looking at the engines. The next location proved even more of a disappointment for Fred. Finch Foundry is an early-eighteenth-century water-powered forge in the village of Sticklepath on the edge of Dartmoor. I'd recced the location and thought it was perfect. Fred would love it. But when we got there for the filming he was less than impressed. First of all it was the men who were working there. For Fred it always had to be the real workers, the men who'd earned their living doing the job they were demonstrating. He never had much time for academics or for demonstrators showing the public how something was done when they'd never done the job themselves. The demonstrators at Finch Foundry were very good at talking about the machinery and showing how it worked, but Fred quickly established that they weren't real forge workers.

But that wasn't all. For Fred there was an even bigger problem. The main piece of machinery in the forge was a big wooden tilt hammer. It was powered by the water-wheel and connected to it by belts and line shafting. Fred took one look at it and said, 'We can't show that. It doesn't work properly.' It looked fine to me but what Fred had noticed was that the belts that drove it were slack and they were not lifting the hammer more than a couple of inches. He said, 'That wouldn't hammer any-thing substantial. You couldn't even get a decent-sized piece of iron under it. It's no wonder they're only doing those teeny bits of things.' Fred wanted to fix it. Before we filmed Fred operating the tilt hammer he'd have to get hold of a ladder and have a go at tightening the belt so that the hammer would lift a bit higher. He had a try but couldn't get much of an improvement, so he went through the motions of operating the hammer as we were filming. When the programme was edited it looked fine and it was unlikely that more than one in a thousand viewers would notice there was anything wrong with the hammer. But for Fred anything like this had to be perfect and for years afterwards he said how embarrassing he found it to have been seen operating something that clearly didn't do the job it was intended for properly.

For Fred the highlight of the first filming trip was saved until the end. On our way back from Devon and Cornwall we were stopping at Bristol to film on SS *Great Britain*. Fred was looking forward to it. Isambard Kingdom Brunel was his hero and in his honour he'd brought his top hat with him to wear on the ship while he was talking about I. K. Fred had quite a few pieces to camera to do on deck, telling the story of Brunel. The trouble was speaking to camera was something that Fred hadn't been able to get the hang of. He was very much a people person and he always had to have somebody he could talk to. The result was that all through the week, whenever he had to say something on camera, he would fix on to my eyeline and speak to me. If I tried to hide behind the camera he would latch on to my daughter Kathryn, who was the production assistant, or Bob Sandy, the researcher, and deliver the line to them. Even if I did manage to start him off speaking to camera, within seconds his eyes would wander away and find me. Fred had to have that human response and it was no different on SS *Great Britain*.

When we got back to base I went over to Bolton with Mark Hamilton, the cameraman. We took some of the things we'd shot to show Fred and let him see that it didn't look right if he was always speaking off camera. We told him how he needed to think about telling the story to one man who was sitting in his living room. Mark said he should think about that man being in the box behind the lens of the camera. 'Always talk to the man in the box,' he said. It wasn't easy for Fred, but from that time on he did get into speaking to the camera. Throughout all the filming I did with him from then on, he would often say questioningly before a take, 'To the man in the box?'

We filmed *Industrial Age* throughout the rest of the summer and early autumn. It was a typical British summer, with a few warm, sunny days, quite a lot of dull, rainy days and some big variations in temperature. One of the first things that struck me about Fred was the way he was completely unaffected by the weather. When it was warm we'd all be in T-shirts and short sleeves; when it rained the waterproofs and umbrellas would be out. But, whatever the weather, Fred would always wear exactly the same things. He never took his jacket and waistcoat off when it was hot and he never put a raincoat on however hard it rained. He always seemed completely oblivious.

Very early on we went to Killhope Lead Mining Centre, high in the North Pennines near Alston. It's one of the wettest places in England and, when we went there to film, the weather was foul. The cloud was low and there was a constant heavy drizzle, the sort that gets into your clothes and has you soaked through in no time. Most of our filming was out in the open. Fred hadn't got a raincoat with him, but it just wasn't a problem. The weather was appropriate because he was talking to the curator about the children who worked at the lead mine in the nineteenth century. They worked outside in all weathers at the hushing tubs, sorting and crushing the lead ore. With Fred standing there in his soaking clothes, it was clear to see the empathy he had with those poor little kids from a century or more earlier. It was something his friend Michael Webber had noticed about Fred from the times he spent with him on his roller. 'He was very peculiar in a way for his interest in people who had hard times; people who worked in a time when there was poverty and people had very little. Fred always managed to relate to people who had little.'

Our industrial tour took us all over Britain but to Fred it was all England. We went to the Scottish Mining Museum near Dalkeith, just south of Edinburgh. The World Cup was still on and while we were there England lost to Argentina in the quarter final, the match in which David Beckham was sent off. We all watched it in the bar of our hotel and Fred, who was becoming a bit of an England fan, couldn't understand why so many of the people in the bar were supporting Argentina. Scottish nationalism wasn't something he understood; in fact he didn't seem to recognize that Scotland existed.

The next morning, when we were filming at the pit, it came up again. There was a long row of semi-derelict Lancashire boilers that had raised the steam to drive the winding engines. They were a sad sight now with weeds and grass growing between them. I asked Fred to walk along the line and tell us how the scene would have looked when the pit was in its heyday. Fred excelled as he painted a wonderfully evocative picture of what it would have been like when the pit was working. It was perfect apart from one small thing. Right at the start of the piece Fred had said, 'This is the finest row of Lancashire boilers I've ever seen in the whole of England here at the Scottish Mining Museum near Edinburgh.' No, Fred. Can we do that again? We're not in England. Can you say the whole of Britain? The second time his description was even better but it was still 'the whole of England'. After a couple more takes I said to him, 'Remember all those fellows in the pub last night? I don't think they'd be very happy to hear you describing this as England.' But it was all to no avail. We did quite a lot of filming in Scotland and Wales over the next few years, but Fred always found it difficult to say Britain or the UK. He was a great patriot but for Fred it was forever England.

Filming in Wales presented its own problems. 'This being a TV presenter has its drawbacks,' Fred used to say, 'because some of these Welsh names are very hard to pronounce.' His favourite was 'Here I am at Llechwedd Slate Caverns near Blaenau Ffestiniog.' I'm sure we tried it a dozen times and whenever Fred got close, there was a problem for camera or for sound. Fred was learning the trials and tribulations of his new role.

One of the things Fred felt strongly about was that, at a lot of the industrial museums and heritage centres we visited, so few of the machines worked properly. 'If you go to the greater part of these museums,' he would say, 'they don't actually do anything with the steam engine. The engine is there or they have an engine that goes round. But then when you go and open the door you see it's not connected to anything so it's very hard for the un-enlightened to see what it did. The closest we can get to what things were really like is Queen Street Mill at Burnley. It's all working there.'

We spent a day with Fred at Queen Street. He stoked the boilers and had a look at the engine but what was particularly memorable was the conversation we filmed with Evelyn, a retired weaver who was a demonstrator there. We had to cut it down to four or five minutes for the programme, but it really needs to be seen in its entirety. It's one of the best examples we have of Fred's remarkable ability to talk to ordinary working people on camera and enable them to be as natural and spontaneous as he was. This was where Fred was at his best. It stemmed from what Hugh Dixon of the National Trust said was 'that extraordinary respect he had for the people and what they were doing. In turn he was respected by them and you had a really good interchange when he visited somewhere, and you felt you were really learning what was going on.'

Fred was very much on home ground when he looked at the textile industry. Introducing the programme for a Dibnah Day on UKTV, he said, 'This next film's about the Lancashire textile industry and the cotton mills and the bleach works that were once in abundance all round here. Both my mother and father spent all their lives working in the bleach works and it's been amazing how much of this stuff has been preserved by really dedicated enthusiasts, like the Northern Mill Engine men with the grand collection of stationary steam engines they have gathered up from all over Lancashire. Some of the machinery is quite unusual. My father worked a thing called the beetling machine which was an unbelievable creation. But I don't think there's anywhere now where a complete mill or bleach works has survived.'

It was this personal knowledge that gave Fred a credibility that few other presenters had. John Yates of English Heritage summed it up. 'Fred's contribution to how we lived was a particular understanding of how the industrial working classes lived and what was expected of them because he was just old enough to have heard old people talking about the age before the welfare state; the age before a real caring about housing quality. I think in his odd little asides he's making clear that the good old days weren't always good but they were heroic.'

As we filmed, Fred started to enjoy it more and more. One of the things he enjoyed most was meeting the people and one man he never forgot was Peter Lewis, who was restoring a windmill in Shropshire. 'I must say amongst all the places we went the standard of his restoration work were magnificent,' said Fred. 'I often wonder if he ever finished it off. He told us about the central spindle in the windmill; how he'd actually turned it out of a great oak tree trunk, using a Land-Rover as motive power. And what we don't see in the programme, this chap actually had his own coal mine in his back garden. I spotted it in a lot of trees. It was a bit like mine; you'd never have known it were there.' It was men like Peter Lewis who really impressed Fred, men who took on big projects and did it all themselves, superbly.

Some of Fred's favourite locations were the big open-air industrial museums. 'When we were doing the *Industrial Age* series,' said Fred, 'we got to all sorts of wonderful places like Ironbridge Gorge and the Black Country Museum.' Glen Laws of the Ironbridge Gorge Museums Trust said, 'We always enjoyed Fred's visits here because he was so knowledgeable about the great age of industry that we try to interpret at Ironbridge. Fred actually knew our ironworks when it was still a functioning ironworks in Bolton. He used to tell us stories about how things actually went on, some of which were very surprising, like the fact that there were broken-down armchairs quite near the furnaces where people would go and relax.'

We went to the Black Country Living Museum with Fred to film their replica Newcomen atmospheric pumping engine. There are no original Newcomen engines still working anywhere so the replica at the Black Country Museum is the only one you can see in action. It's quite a temperamental thing and when we got there to film it things started to go wrong, as Ian Walden, the director of the museum, recalls. 'We'd come in early to make sure it was working all right. We got it started but it was a little bit awkward to run and in trying to run it we managed to detach the chain from the beam on the outside of the engine just as Fred arrived. I turned round to him and said, "Look, we need somebody who can go up a ladder with a chain block over his shoulder and connect those bits together." I'd said it a bit tongue in cheek but he looked at me straight in the face and said, "Well, find us a ladder, lad." Ten minutes later he was up there fixing it and he had the engine running. In another ten minutes, he was back into filming and performing for his public – absolutely great bloke.'

On the domestic front, Fred and Sheila had decided to get married. Sheila had moved in with Fred but he was quite old-fashioned in his views and wasn't comfortable with this modern way of living together. So a wedding it had to be. The big event was planned for September, towards the end of our filming schedule. Before the wedding, Sheila joined Fred and travelled round the country with us on the last big trip we had planned. What impressed her right from the start was the way Fred would have a go at any job when we were filming. 'He did appreciate skills,' she said, 'and I think the reason was because he was able to apply many of those skills himself. He could turn his hands to anything. One of the places I most remember going to and realizing just how much skill Fred had was when we went to the Welsh Slate Mining Museum at Llanberis, and Fred had a go at slate-splitting and trimming. The boys who were demonstrating there had been doing the job for many years and they made it look very easy. I thought to myself, Fred's not going to be able to do this. But, enthusiastic as ever, he sat down and took a piece of iron and chop, chop, chop, he started. He did a perfect job and I think everybody was amazed that he managed to do it. According to one of the men there, the apprenticeship for the job was something like five years before you were allowed to do what Fred had just done.'

In the pub that night Fred met another famous climber who was based in Llanberis. Joe Brown was the Manchester plumber who, in the 1960s, had pushed forward the boundaries of rock climbing and pioneered many new routes, particularly in Snowdonia. He came over to talk to Fred and the two great climbers compared notes about their very different types of climbing. Joe said, 'I couldn't do what you do, climbing up vertical ladders like that with no ropes and nothing to clip yourself on with.' Fred in his turn said, 'There's no way I could do what you do on that rock. It all looks too dangerous and unstable for me.' There was clearly a lot of mutual respect.

As we were going round, Sheila encouraged Fred to spend more time looking at his script notes and trying to remember a bit more of what he was supposed to be saying. But there were far too many other things that caught Fred's eye. I always gave him a set of notes to look at as we were driving to a location, but they would soon be cast aside as Fred would dig me in the arm to point out an interesting chimney or a railway viaduct. And it wasn't any better when we arrived, as he always found somebody he wanted to talk to. At Kew Bridge Steam Museum we were setting up a shot where Fred would be talking about one of the big pumping engines there. I went through the script with him. What he needed to say was, 'This engine pumped over a million gallons a day and it provided water for the whole of West London.' I left him to look at the notes, but within a minute he was deep in conversation about the engine with one of the volunteers from the museum. He called me over excitedly and said, 'Do you know, this man's just told me that this engine pumped over a million gallons a day and provided water for the whole of West London.'

A couple of weeks after that filming trip Fred married Sheila. The reception was at the Pack Horse Hotel in Bolton. For me it was a day off, but Nick Wilding was there with his camera filming the whole event. There was a band and Eddy Chatwood's daughter was up there singing with them, but it wasn't really Fred's style of music. And he was no more of a dancer now than he had been as a teenager. I sat talking to Fred with my wife,

Fran, and his old mate Derek Roscoe. I went to the bar to get some drinks and when I got back a girl in her late teens had joined the group and Fred was in tears. It was his youngest daughter, Caroline, whom he'd not seen since she left with Alison when she was still a toddler. He was upset because when she came up to him he hadn't even recognized her.

Fred said he had invited all his daughters to the wedding reception but he was surprised that one of them had turned up. After all, they'd not seen much of their dad or heard much from him for a long time. By his own admission Fred hadn't paid any attention to his daughters after they had left with their mother. None of them had moved away from Bolton but there had been little or no contact. It was different with his boys. Even though they lived in the Isle of Man he still saw them whenever he could, but for Fred, with his old-fashioned, patriarchal values, girls were not the same. Over the years we worked together Fred would talk regularly about Jack and Roger but never mentioned Jayne, Lorna or Caroline.

After the wedding there was still a little bit of filming to do. We went with Fred to look at his restoration work at Wetheriggs and Glynllifon and the chimney he'd done recently at Barnoldswick. There were also some stories for Fred to tell about Bolton and the influence that the town's industrial landscape had on him, and there were the shots to be filmed of him driving his Land-Rover. So how had he found the filming? 'In the old days or when I first started,' he said, 'it were very different to what it is now. I didn't have to talk to the camera, number one; I found that rather strange talking to a window sort of style. They always did voiceovers and had a narrator to do the talking; you know, I just did the action stuff like Clint Eastwood.'

Fred's screen persona from those earlier series hadn't done much for Sheila. 'I remember seeing this horribly oily man on the telly, dragging his poor wife along and telling her to take the kids off for some chips. I was watching with my parents, and I said no wonder everyone thinks we all wear clogs in the north.' But she felt there had now been a significant change to Fred's image. 'Fred was known just for his cheeky chappy grin on television until the filming of *Industrial Age*. After that people tended to take him more seriously because they saw quite a depth of character there. He was very passionate about things from the industrial past and that passion made more viewers familiar with it. People could see it wasn't phoney with Fred; everything that you saw in him was genuine. Through that series and the ones that followed Fred did a lot to raise public awareness of industrial history. No doubt about it, before Fred came on television, there didn't seem that much that was devoted to industrial heritage. But if you mentioned industrial history to someone in the street now, the person they would most readily associate with it would be Fred Dibnah.'

Sheila found that Fred was now an even bigger celebrity and they couldn't go anywhere without him being instantly recognized. 'Fred had a great ability to make friends with people that he'd meet. We'd just go into a pub or stop off for a meal on the way back home from a rally, and immediately somebody would come over and start talking to him. Fred was always very pleasant with everybody. He would have some really in-depth conversations with complete strangers and would make a lot of friends in that way. He'd exchange telephone numbers and this person would ring him up a few days later. Then they'd get the standard invitation from Fred, "Why don't you come down and have a look at my yard?"'

The first steam rally that Sheila had gone to with Fred was the one where the organizers had provided the tent with the two bus seats for them to sleep on. But by now she had achieved a significant upgrade in the kind of overnight accommodation Fred was offered when making personal appearances. He was being invited to more and more events, ranging from steam rallies to black-tie dinners, and Sheila would often accompany him. But the couple never shared anything as humdrum as a holiday – owing to Fred's complete lack of enthusiasm for putting his feet up. Some things never changed.

9

Magnificent Monuments

Fred was back on television, only now he was there in a different guise. It was still the same old Fred in many ways, but he'd made the transition from being the subject of documentaries to standing in front of the camera as the presenter. And the transition had been a great success. Fred's old fans had all switched on for *Industrial Age*, so viewing figures were high. The website that Fred had promoted so self-consciously in the programmes was the BBC's second most visited site at the time, beaten only by the Teletubbies. The book that accompanied the series made it into the top ten best-sellers' list and was one of the top five history best-sellers of the year. Fred was back on the map and so was industrial history, which had, until this time, been a bit of a backwater.

'Affectionate' and 'unpretentious' were the words most used in the reviews of the series: 'A series paying affectionate homage to Britain's industrial past', 'Dibnah's affectionate series, without minimizing the hardships and dangers of heavy industry, is a poignant tribute to the communities that grew up around it', 'A sweet piece of heritage television. An unpretentious introduction to industrial archaeology.' Fred himself was described as 'a regular Indiana Jones when it comes to unearthing the artifacts of lost civilizations such as that known as "Industrial Britain". He drools over old beam engines and water mills that have been saved from the scrap yard and returned to their former state.'

Fred had learned a lot about presenting in that first series, but he hadn't found it easy. Remembering lines and saying them to camera had been difficult. Often there were five or six takes, sometimes more, to get something close to what needed to be said. Names and dates were a problem and the order in which things happened and, although Fred liked being the centre of attention, he was very self-conscious. This was aggravated by the fact that in many locations a small crowd would gather round to watch as we were filming. After a few takes he would turn to them with a grin and say, 'I'm not a bloody actor, you know.'

In press interviews after that first filming Fred admitted that being a TV presenter was more nerve-racking than climbing a three-hundred-foot chimney. 'I'm all right if someone asks me a question, but I find the mechanics of making television unnerving. I don't mind talking – anyone who knows me will tell you that.

But standing in front of a camera is a different matter. I enjoy the TV stuff, but I'm much better at the steeplejacking.'

But the success of *Industrial Age* meant we could develop other ideas for series that Fred would present. When I talked to him he was keen. He'd enjoyed the filming because it had allowed him to travel around the country to see lots of things he was interested in. So what was the next series going to be?

Given Fred's passion for the subject, steam was an obvious choice. As far as I was aware, nobody had ever done a really good history of steam power and of the various applications of steam. We researched and developed a number of proposals, but there was going to be quite a big overlap with the series we had just made. Ideally we needed a different subject that Fred was equally passionate about. His interests were wide-ranging, from breweries to weapons, but when I talked to him one thing that kept coming up was buildings – not so much architectural history as how places had been built.

'I first became interested in buildings,' he told me, 'when I were about sixteen years old. My work as a joiner got me into some of the splendid mansions that the cotton mill owners or bleach works owners had built. I couldn't help but notice, coming from a house that hadn't got any skirting boards, the quality of the woodwork and the height of the skirting boards, eighteen inches up the walls, and the beautiful panelled doors and architraves and best of all the fancy plastered ceilings. It really intrigued me as to how it were all done and I spent from being sixteen till I were twenty-two doing that sort of work, which I really enjoyed. It taught me a lot about buildings.'

He also told me about one of his own building projects. It had been inspired by the chimneys he remembered from when he was a child. 'When I was a kid all the beautiful chimneys in Bolton had wonderful names. This is because in the olden days, when a bunch of men came together as directors to build a spinning mill, one of the first things they did was to christen it. Then, when they'd built it, they'd put the name on the chimney in white tiles. So, when I was a little lad, Bolton was surrounded with them. They all had names like the Dove, the Ocean and the Swan and they all had beautiful tops, great flared fancy overhanging ones. But as all the chimneys in Bolton have disappeared and as my life has progressed on to my twilight years, as you might say, I'm pleased to say I've been able to build a chimney myself in my back garden. It's nearly finished and I'm going to have a topping-out ceremony like they did in the olden days. There are many tales about the finishing off of great buildings and chimneys and they always had a topping-out ceremony. So I'm going to take my old gramophone to the top and take a crate of Guinness up with me and I am going to have my own music and a pint on top of the chimney that I have built in my own back garden.'

It seemed like a perfect opening for a programme on building and on the preservation of historic buildings.

Fred clearly had a long history that influenced him. A lot of people knew him as someone who knocked chimneys down, but he had been involved in the conservation of chimneys as well. Even when he did do a felling, because of the way he knocked them down, he actually had to know about how people had built them in the first place.

The proposal that came out of this was for *Fred Dibnah's Magnificent Monuments*, a series in which Fred would give his own distinctive view of some of the country's most famous and historic buildings – castles, cathedrals, abbeys, great houses and engineering marvels. The starting point would be Fred's own experience and he would call on that to give answers to some of the questions millions of people ask when they visit a historic building: 'How the heck did they build that all that time ago? How did they manage to lift all that stone and wood up to such a great height?' Fred would travel around Britain to look at the skills and techniques of architects, civil engineers, builders and construction workers throughout our history and at the craftsmanship of the carpenters, wood carvers, stonemasons, blacksmiths and glass makers whose work adorns their buildings. He would tell us how they overcame the huge problems they faced and the construction techniques they used; the time they took and the amount of manpower that was needed. He would reveal some of their ingenious methods, the sort of machinery they used and how it was powered. The series wasn't going to be about kings and queens and the people who lived and worshipped in the places we were to visit. It would be about the ordinary working men who built them.

It was a different approach to history and one that Fred was very keen on. 'That's the kind of history that really grabs me,' he said. 'So often you go to these stately homes or wherever and the guides that are showing you round have got it all off pat. There is no enthusiasm, no excitement, no oomph! Ask them a question about how the place was built that's not in the script and they dry up.'

Fred also liked the idea of doing a series about magnificent buildings because it would redress the balance a bit and show a different side to him. The image people had of him as a destroyer of buildings was not one he was particularly happy with, as Sheila remembered. 'He'd started off, as a young man, watching the steeplejacks climbing these chimneys and repairing them, and when he had to demolish them he was very sad. He'd get photographs out of past jobs and say, "Ay, well, that's no longer there. I remember a time on the golden skyline when you could see dozens and dozens of these things." There was a great sadness in Fred when he had to demolish so many chimneys, so to be seen to be involved with the conservation and preservation of buildings was good for him.'

By the time *Fred Dibnah's Magnificent Monuments* was commissioned by the BBC, Fred's life had changed and things were good for him again. Financially he was getting back on his feet and he was settled down with a new wife – 'the current Mrs Dibnah', as he always referred to Sheila. After the success of the first book, another was commissioned to accompany this second series and by this time Fred had got a new agent who was getting him more after-dinner speaking engagements. The steeplejacking business was tailing off, but all this new work was adding to the revenue stream. There were some minor problems he had to put up with, like the fact that Sheila was vegetarian and would only cook vegetarian dishes at home. 'But,' he said, 'she's a bloody good cook and some of this stuff tastes really good.' And when he was away on filming trips he could always go back to what he regarded as proper food. All the more reason, therefore, to do some more filming or to join me on the research trips for the next series.

These research trips for *Magnificent Monuments* were some of the happiest times with Fred. By day we were visiting some of Britain's best castles, cathedrals and stately homes and clambering over some of our greatest civil engineering structures. Not only that but we were getting privileged access to the parts of them that the general public never got to see, inside the walls or up on the roof. Fred hadn't been to many of these places. Visiting castles and cathedrals was something that people did when they were on holiday or having days off, and those things had never fitted into Fred's timetable. But he was fascinated by what he saw. What became clear very early on was the ability he had to spot where things had been altered or added to over the ages; where a line of brickwork changed, where a window or a door had been blocked up or where a wall had been added.

When we went to the Tower of London Fred was fascinated by the building material used by the Normans in the White Tower. He decided that it was like an early form of concrete. 'It's only the mortar that holds the thing together,' he observed, 'because, although there are corners and things where they have used nicely cut blocks of stone, a heck of a lot of it is just rubble. They were damned good at making mortar.' And straight away he related it to his own efforts. 'When I look at the way that it was built and the length of time that it has stood there, it gives me hope for my building work.'

Fred loved meeting people and talking to people and that was one of the things that gave him so much enjoyment from the research trip and the filming that followed. 'During the course of these trips,' he said, 'we got to meet a lot of important people like surveyors and clerks of works. When we went to St Paul's Cathedral we met the gentleman who's in charge of the whole place. His name is Martin Fletcher and he is actually a joiner by trade. He was born in London within sight of the dome, and in the vaults of the cathedral they have a great plaque on the wall with the name of each man who's been in charge of it right from the days of Sir Christopher Wren. Martin Fletcher was very proud that his name was the latest to be carved on that stone and he told us how proud his mother had been when she came to the ceremony when his name was unveiled, especially as he'd started as a humble joiner just like me.'

While we were at St Paul's Martin took us out on to

some scaffolding around the outside of the dome. When you get to see it close up, there's a lot of evidence of repairs done in days gone by, but we had been taken out to see some really impressive restoration work that was being done for the first time. It revealed some of Wren's engineering work that keeps the dome standing. Inside the stonework, there's a big wrought-iron band that goes all round the base of the dome where it meets the main roof of the cathedral. Over the years damp had started to seep through the stone and corrode the metal. The metal had then expanded and had begun to crack the stonework, so the metal band and the stone were being replaced.

Fred loved it out there on the scaffolding. It was his natural habitat – a building site up in the sky. But he had to be dragged away. While we were in London we had been invited to lunch by BBC Books to celebrate the success of our *Industrial Age* book. We got a taxi from St Paul's to the restaurant and as we stood outside we looked at ourselves and the state we were in. We looked like a couple of construction workers who had just walked off a building site. The work on the dome had involved a lot of stone-cutting. It was white Portland stone so we were covered from head to toe in fine white dust. There was nowhere we could go to get changed and tidied up so we marched into the restaurant. We were a few minutes early and our hosts hadn't arrived, but when we said we had a table booked by the BBC, the maître d' showed us to our table without batting an eyelid.

After lunch we went back to St Paul's to continue our

tour with Martin Fletcher. We talked to him about the construction of the dome and he gave Fred a wonderful cutaway drawing showing all the detail of the structure. That evening back at our B & B, we looked at the drawing and I said we could do with something like it for the programme so that Fred could explain how Wren had built the dome. 'I'll do one,' he said. 'I'm quite good at drawing, you know.'

One of the people who knew just how good Fred's drawings were was his friend Alan McEwen. 'His drawings were beautiful, absolutely superb. He could set to with a pen and ink and a decent piece of paper and do a drawing virtually anywhere. I appreciate this because I do lots of really serious drawings for my company, yet I have to be in the right place to do it. I've got my drawing board up; I've no clutter, no noise, no bothering me. But Fred could work in a cacophony of noise, and people bothering him and mithering him, and he'd be able to do a magnificent drawing.'

When Fred got home he started work straight away on his sectional drawing of the dome of St Paul's Cathedral. After a few days he called me to say it was finished. I went to Bolton to have a look and it was superb – even better than the original that he had been given at St Paul's. It was the first of many drawings that Fred was to do, which would add an extra dimension to the programmes, and for me that first one was always the best. By this time the programme content and all the locations had been decided on, so I selected a number of key structural features from them and asked Fred to do further drawings to illustrate them.

While I was at Fred's that day I was sitting at his kitchen table reading his *Daily Mail*. One of the headlines was about Alex Ferguson, the Manchester United manager, giving David Beckham a dressing-down for missing a training session. We'd still not turned Fred into a football fan but I called him over to have a look. Beckham had been spotted in London's most fashionable restaurant, the Ivy, when he should have been training in Manchester. Little did we know of its status at the time but that was where we'd had lunch a few days earlier, when we'd come from St Paul's looking like a couple of builders.

Before we started filming one or two changes needed to be made. To capture Fred's enthusiasm and spontaneity I wanted a crew who would be able to work quickly and start shooting very soon after we arrived on a location. Mark Hamilton was an excellent cameraman who had a particular skill for lighting things. The trouble with good lighting is that it takes time, which meant there had been occasions in *Industrial Age* when Fred's first flush of enthusiasm had died down a bit by the time we were ready to film. Andrew Hartley, who came in to work on the second series, was much more used to shooting things with less set-up time. Once we started shooting he and sound recordist Nigel Chatters developed a good rapport with Fred and that crew was to stay together for most of the films we made.

Our first week's filming took us to Hadrian's Wall and then over the border into Scotland. Straight away Fred demonstrated his own down-to-earth slant on history as he embellished the script notes that had been written for him in his own unique way. 'They worked bloody hard on Hadrian's Wall,' he said in the voice of a man who knew all about the construction game. Then he gave a graphic description of the workings of the latrines at Housesteads Fort, finishing off by telling us that the 'affluent' went off into the fields, which would have 'made it all a bit stinky down there'. The next day, standing below Edinburgh Castle, he did his scripted piece about the way the castle had been built on the rock, then as he looked up at the craggy rampart he added, 'By heck, you can imagine the enemy looking up at that lot and saying sod it, let's run.' It was these little asides that made Fred so different.

Everywhere we filmed, Fred was able to relate what he saw to his own personal experiences of working in high places and getting himself and his building materials up spires, towers and chimneys and on to rooftops. Because he'd been there and done it all himself he could speak with authority about the practicalities of building massive structures and he showed a real appreciation for the scale of the achievements of the builders of the past.

By now Fred was not doing much steeplejacking but he'd always liked church steeples, and there was one left that he still wanted to do. Rising to 309 feet, the church of St Walburge in Preston has the third highest steeple in

Britain and it was in need of urgent attention. Fred had already laddered the steeple and done some preliminary inspection work. It was going to be a big job, but with all the filming he was doing now, and his speaking commitments, there had been a lot of demands on his time.

When we came to film *Magnificent Monuments* Fred's red ladders were still on St Walburge's steeple. We needed an introduction to the programme on churches and cathedrals so we filmed Fred as he climbed to the top of it. For the viewers Fred was able to call on all of his steeplejacking experience to explain how church steeples like this were built and relate it to the work that needed to be done on St Walburge's. 'When people see the enormous thickness of the walls at the base of a church tower,' he said, 'they have this strange idea that the steeple is almost solid, but it isn't. A steeple is very thin and very light. So why doesn't the top blow off a church steeple in a gale? The answer is that it has a big nut and bolt on the inside that comes down for maybe twenty-five to thirty feet. The nut is screwed on at the bottom, and that has the effect of clamping the capstone down to the top twenty feet of masonry. That's why the weathervane stays up there in a gale, and doesn't end up in the cemetery. But as the years go by, the rain gets in at the top and runs down the rod and the rod starts to go rusty. Some church steeples have access to get at this rod, so that you can paint it. This big steeple in Preston has got this very thing. It has four cast-iron, Gothic-arched opening lights that have never opened for years and they have that awful, half-inch-thick, dark green glass in them. Part of the glass is broken, so you can see the rod inside and see that it is in quite good condition. The top thirty feet of this steeple is only about two feet six inches across. It is really fine, just like a needle; so fine that the top of the steeple rocks in the wind. Because of all the rocking, the head joints have cracked. Now, if this was a chimney stack, the obvious thing to do would be to put great big iron bands around it, like a set of corsets. There are some big steeples that have a lot of ironwork on them, but I think it would look rather unsightly on this one, and I think I can get away with less drastic methods of keeping the thing up. First of all, though, I'm going to have to do an inspection, which entails dangling three hundred feet up in the sky and leaning outwards on the end of a rope to look at the ornamental bits, and see if there is anything ready for falling off.'

But it wasn't to be. Fred never found the time to return to St Walburge's and I believe the day we filmed there with him was the last time Fred ever climbed a steeple. He didn't even go back for his famous red ladders. They stayed there for a few years until the parish priest called in another steeplejacking firm to do the job. Fred realized that he wasn't ever going to need his ladders again so he donated them to the next man.

Just as Fred was able to use his steeplejacking knowledge to tell us how things were built, he was also able to call on his experience as a joiner when it came to talking about the interiors of the places we visited. He'd worked with wood when he was doing his joinery apprenticeship and he could speak with authority about the work of carpenters and woodcarvers and give us a real appreciation of their skills. Wherever he went where restoration work was being done, he enjoyed talking to the craftsmen and women, looking at what they were doing and finding out from them what the original techniques had been and how they went about copying them.

At York Minster, Fred was very impressed by the work of the stonemasons. 'York Minster,' he said, 'has to be, without any shadow of doubt, one continuous effort that will go on for ever just to keep the thing standing up. I bet they have been working on it every day since it was built. People say it is a pity that there is nobody with the skills to do this sort of thing now, but round the back of the Minster there's a good collection of craftsmen and women all beavering away. There was a young lady there who was making the most beautiful finial. I had a good chat with her and asked her how she went about starting it off and, as a result, if I were asked to make one now, I would have a much better idea of how to go about it. She had been chiselling away at this one particular piece for eleven weeks. Just think about it, every day for eleven weeks for one piece and it was only about two feet tall. But think of that wonderful feeling of satisfaction when you get to the end of it, when you consider that you started with a pyramid-shaped lump of rock, two feet tall, eighteen inches at the bottom and one inch at the top, and then you transformed it into a beautiful thing.'

Whatever the work was, Fred would always have a go himself. Particularly memorable for him was Ightham Mote in Kent. 'It was one of the most beautiful places we went to,' he recalled. 'It was a lovely half-timbered, wattle-and-daub construction, which was in the process of being renovated. They had me doing a bit of mixing the cow dung which made the daub to stick on the wattle, as you might say. It was a bit stinky, but it was all good fun.'

Hugh Dixon of the National Trust spoke about this side of Fred's work. 'Fred was very good at seeing what was going on and he often dealt with topical things, like the conservation we were doing at Ightham Mote. What was particularly good was the way he looked at how the walls had been constructed originally, and how much trouble had been taken to restore them in the original way, using exactly the same materials. That was genuine conservation.'

Another memorable visit was to Richard Rogers's magnificent stainless steel and glass palace in the City of London, the Lloyd's Building – home of the insurance men. Fred wasn't too keen on modern architecture but he was glad we were going to Lloyd's because he was having some problems with his steeplejacking insurance and he thought he might be able to get a good deal from someone there. 'In the early days of my steeplejacking career,' he explained, 'the insurance weren't too bad. I never had any claims and the premiums were quite low. But as the years rolled by, the premiums for chimney-fellings went up because of all the damage that the dynamite men did. When we turned up at the Lloyd's Building, it were the architecture that we were concerned with, not insurance, but we were on the trading floor and there they all were with their pinstripe suits on and all taking the mickey out of me, saying things like "What have you come to blow up, Fred?" and all the rest of it. But as soon as I tried to make some serious enquiries about steeplejacking insurance, everybody disappeared. They all ran like rats down a sewer, and I couldn't get anyone to discuss it with me or give me a quote. In fact I've given it up now altogether. It's got to the stage where they want more money for the premium than you get for doing the job.'

Fred didn't like London much, but he did like a lot of the places we went to and the people we met. As a joiner, one place he had a particular appreciation for was the timber-framed Globe Theatre. 'We went to this lovely Shakespearean theatre,' he said. 'It was all made of oak and built by a man called Peter McCurdy. One of the interesting things I found out from him was that the two great pillars on the stage of the Globe Theatre have got a hole right up the middle of them. And Mr McCurdy explained how they had to bore this hole down the middle of a tree about twenty-five feet long.' Fred was fascinated by the Globe and greatly respected Peter McCurdy's craftsmanship and wealth of knowledge about timber-frame construction. It was a world away from steam and industry, but Fred's enthusiasm was just as strong.

One of the things Fred particularly liked about the filming of *Magnificent Monuments* was the great range of structures we looked at. 'What an unbelievable contrast from Shakespeare to the Pepsi Max Big One at Blackpool, which we filmed a few weeks later,' he said. 'What a lot of people don't know is that I had to ride on this thing about three or four times to get the film made. What made it worse was that it was early in the morning, before the ride had opened to the public, when everything's cold and horrible. The wheels are made out of some sort of plastic material and first thing in the morning they are very hard so the ride is incredibly bumpy. You have to go round about four or five times before the wheels have got to the right sort of rubbery consistency and it's a much smoother ride then. They didn't tell me that till after, though.'

Filming with Fred got us to some splendid high places, like the top of both the Forth Railway Bridge and the Road Bridge, the top of one of the towers on the Humber Bridge, the top of Big Ben and the lantern of Ely Cathedral. None of these places was dangerous because there were always railings and plenty of protection and as long as you had a head for heights you never felt too exposed. But I was always grateful that, by the time I did my filming with Fred, his steeplejacking days were virtually over. If they'd not been I know he would have insisted that I went up to the top of a few two-hundred-foot chimneys with him. I couldn't have said no, but I don't think I would have been very happy about it.

For *Magnificent Monuments*, in order to save Fred from having to remember too much scripted material we'd decided sometimes to film him being shown round by guides. In this way we thought we'd be able to get a lot of the historical background from them. But this turned out to be a bit of a problem because Fred wouldn't let them get a word in edgeways. Fred was a great talker but he wasn't always the world's best listener. He had a wonderful habit of asking a question and then answering it himself, before his interviewee had a chance to say anything, or cutting into what they were saying with one of his own stories, which made the editing a little difficult.

In all the filming I ever did with Fred there was only one man who managed to silence him completely. It was at Conwy Castle and Fred never forgot the day he was rendered speechless. 'We had a guide,' he said, 'quite a small man, but he had a walking stick which he lashed about in an unbelievable wild manner. Just standing near this chap was quite dangerous because of the way he flung his stick about.' The guide was clearly a great enthusiast but I had to cut the whole sequence because these were Fred's programmes and the guide took over the show.

On our way back from the pub in Conwy that night we had a little run-in with the local constabulary. We'd had another lock-in so it was quite late when we were walking back to our lodgings just outside the town walls. As we walked past one of the medieval towers Fred noticed there was a great big crack running down it. He had to have a closer look. We worked out that the tower was directly above a railway tunnel and Fred reckoned the crack must have been caused when Robert Stephenson was building the railway line in the 1850s. It was a very good illustration, we thought, of the sort of damage done by the nineteenth-century railway builders and we decided it would be worth doing a little piece about it the next day.

It was one o'clock in the morning but Fred went to have a closer look at the tower to see how far the crack stretched and I started to look at it from different angles to see how we would shoot it. As our moonlit recce continued a police car drew up. My explanation that we were planning some filming for the next day failed to convince, and when Fred came to speak to them the Welsh policemen turned out to be almost the only people I ever met who didn't recognize him. I don't know what they thought we were planning but they were very suspicious. Fortunately our lodgings were only about fifty yards away, but they wouldn't leave until they saw us safely back in there and away from any potential damage we might have done to Edward I's town walls.

All our filming was in tourist locations that had lots of visitors and Fred was very aware of the problems this caused. 'We got to some beautiful cathedrals and castles but the big drawback was the general public always wanted to join in so it took twice as long. There's all the good wishers and the well-doers come saying hello so there have to be lots of cuts. Sometimes the camera is quite a long way from me and they just come up to me when I'm in the middle of trying to remember what I'm supposed to say. If you get short-tempered like these proper actors do and say go away or throw a tantrum and start swearing, you're not a very nice person so I never do that. I've never been like that myself. It's amazing, though, how many of these bystanders would come up to us and say, "Have you come to blow it up?" and things like that. They always associate me with destruction but what I'm really interested in is the way these things were put up.'

Fred had done quite a lot of building work himself and just as we were able to show his industrial heritage preservation work on the engines at Wetheriggs and Glynllifon and the chimney at Barnoldswick in *Industrial Age*, *Magnificent Monuments* gave us the opportunity to show some of Fred's own building work and look at the way he went about it. While he was doing this he gave his own distinctive take on the way some of our greatest historic buildings should be preserved. 'Thirty years ago,' he said, 'I bought the detached cottage or gatehouse that I live in off the Earl of Bradford. It was never really big enough for me, so, just as the castles and wonderful buildings we have been to have all been extended by the men who lived in them, I decided I would build an extension. Basically all these men who owned these places were just like me, all DIY men. Even kings were DIY men with all the messing about they did with their castles and palaces.

'There had been extensions done on my house in the days of the Earl of Bradford, but the original bits had got some nice stone mouldings and Gothic arch windows with moulding round their edges. The work that had been done for the earl had been done very crudely and who-ever did it had not extended the ornamentation into the stonework. So I decided that I would cut twelve blocks of stone to the thickness of the moulding and then put some ornamentation on the front of it. I put all this nice stonework on to the front of the house, but when I had finished it all looked very clean and didn't match. So I got a paintbrush and a tin can and mixed some soil out of the garden with mud and water and painted it all over the new mouldings and let the rain wash it off. Now you can hardly tell the difference; it looks almost the same as the old stuff.'

This was Fred's approach to restoration work and it is one that many people would agree with. But during our filming he found out that his attitude didn't fall in line with a lot of official conservation policy. 'What I did with my stonework is something they could have done with doing on the Scott Monument in Edinburgh,' Fred said when we filmed there. 'They have done a lot of restoration work on it recently, repairing the stonework that had been broken away. It's beautiful stonework but the problem is that all the new stuff they've done has been left perfectly white. It is exactly the same stone that the thing is built out of, but it will never ever go the same colour as the original stone because there will never be that amount of coal smoke in Edinburgh again. I think I would have secured the services of someone like me, on the end of a big piece of rope with a tin of matt black paint, and I would have painted all the new stonework with the matt black paint to make it blend in with the old in the way that I have done with the mouldings on my house. I am certainly in agreement with them on one thing, though, in that they haven't cleaned the monument. A lot of that shot-blasting and cleaning that is done now, apart from making things look like brand new biscuits, can also be harmful to the stone itself.'

Fred was pleased that his new career as a presenter was showing a different side to him. 'In the early days of these films people got me down as being a destroyer of buildings and I never really liked doing that. I like knowing how they were built and building things myself. Back in my steeplejacking days when I was repairing the chimney stack at Cheddars Lane sewage pumping station in Cambridge we did a small sortie into the town itself and around the university to look at the beautiful buildings, and I never really thought at that time that I'd end up doing programmes about buildings like this. As a steeple and chimney repairer I was able, by pure observation, to learn how these things were built and how they went about it. So the furtherance of this, doing these programmes about architecture, were quite a pleasure and it were quite easy for me to stand there talking about church steeples and groined roofs and things of that nature.'

One of the wonderful things for Fred about making these films was the fine craftsmanship we saw in many of the places we visited. A theme that began to emerge was that 'the age of great craftsmanship is not dead'. Many people say, 'They can't produce work of that quality today,' and Fred was particularly keen to show that this was not the case. 'Going around on our travels,' he said, 'we've met craftsmen and craftswomen who are still capable of doing really good-quality work. It's so annoying when you meet people who say we can't do that now. It's really an insult to the craftsmen and the crafts ladies of England. The only great difference is time. In medieval times a man got out of his bed in a morning and he'd not got a lot to worry about other than how sharp his chisel were and his hammer and a big block of stone and his dinner. Of course now everything's about money and time. In the olden days a man could have spent all his life working on a cathedral and now everything's got to be done in a matter of months. Everything's reckoned up in time and expense, which is a bit sad, because to do a really good job you've got to spend the time.'

10

Victorian Heroes

Fred was, by his own admission, a man born out of his time. His era, he always used to say, should have been the nineteenth century. It was the age he admired more than any other, the time he would have liked to live and work in. His heroes were the great Victorian engineers – Brunel, George and Robert Stephenson, Sir William Armstrong. He liked them for their courage, their vision and their ingenuity and because most of them were colourful, larger-than-life characters who had confidence in their ability to overcome any obstacle. They were men for whom no challenge was too great, whose 'tackle anything' approach was one that Fred shared.

January 2001 was the centenary of the death of Queen Victoria and to mark the occasion the BBC were doing a Victorian season. Given Fred's great love of the Victorian Age, it was appropriate that the next film we did with him was *Fred Dibnah's Victorian Heroes*. He had a great empathy and connection with Victorian workers and engineers. For Fred these people might have been tired after working long hours but they still put beauty and quality into what they were doing and it was those values that he connected with. His own house was full of the ornamentation, the ironwork and the bric-a-brac of the period. In the workshop in his back garden, he used many of the engineering techniques pioneered by his Victorian forebears over a hundred years earlier, and all his machinery was driven by the force that powered the age – steam. Above all, many of his opinions and attitudes were those of the Victorian Age.

Fred was very patriotic and he took great pride in the fact that in Queen Victoria's reign Britain was the best in the world at making and inventing things, and the skills of mechanics and engineers were highly valued. As Fred worked on the traction engine in his back garden he showed us that the tools he was using were the same as those employed in a Victorian workshop. Engines were built in little workshops very like Fred's and he demonstrated how iron was shaped for locomotive manufacture using the steam power and the machinery of the period. There are not many people around who can do this sort of thing now, but Fred's view was that you can do anything if you put your mind to it. That was what he liked about the Victorian Age. It was an age of men who were individualistic and entrepreneurial. Their stories were all told at the time in the pages of the *Engineer* and Fred's bound volumes of Victorian editions of the magazine were a constant source of reference for him.

It was an age that inspired Fred. 'One of the virtues of the Victorian age,' he said, 'were the business of self-help. If you worked down the pit or in the foundry, you could stay there for the whole of your life if you wanted to. There weren't much chance of them going out of business like they do now. The thing is that if you wanted to get out, there were such places as the mechanics' institute you could go to where there were books and you could learn how to read and write. If you showed your mettle, what you were capable of doing, Victorian society admired them men and made them into heroes like Brunel. Weaknesses of any sort were frowned upon and you'd got to keep going at all costs. The skills of making things in Victorian times were highly valued and I think somehow or other we've lost all of that. Us English now are not best prepared to get our hands dirty like we used to and other nations have taken that over.'

One of the things that excited Fred was the way in which the British civil engineering industry of the day rose to the challenge of building the railway network. The technical achievements of Isambard Kingdom Brunel and other engineers like Robert Stephenson astounded their contemporaries and set new standards in civil engineering. It was the personal qualities that Brunel displayed that made him such a hero and inspiration for Fred and there is much about Brunel that those who knew Fred well would recognize in him. Brunel was a man of great drive and ambition who worked long hours and had frequent absences from his home and family. He was outwardly confident but inside he was deeply insecure. When we looked at his diaries with Fred, they revealed that Brunel worried all the time that his grand projects would fail. Like Fred, he was a perfectionist whose standards were of the highest order.

For Fred, Brunel represented the values of his age – hard work and innovation. Manufacturing skills were highly thought of; there was little sympathy for weakness and none at all for idleness and, above all, there was an unbending faith in the ability to get things done, however great the challenge. They are all values which Fred thought we'd lost, but if we still had them today he fervently believed that Britain could again be a major industrial power. Fred would always talk of the gradual decline of British industry that came with the loss of these values; the complacency that set in from an early stage and the almost total disappearance of people with his kind of skills.

Ken Moss of the Victorian Society talked about Fred's enthusiasm for the Victorians. 'He liked their technology. He liked their values. He liked the craftsman approach to things. And I think that if you went back to Victorian England, you would have found a lot of Freds around. I think he would have fitted in very well. Fred told it as it was. You could see that Fred was proud to be British and he was proud of the achievements of Great Britain when it ruled the industrial world. I think that Britain has probably contributed as much – if not more – than any other nation to progress in the world. So it's good that Fred was able to say that, and not only say it but to say it with pride and enthusiasm. In recent years the sense that Britain was a world empire and the workshop of the world has become tainted with a sense of imperialism, but Fred always sought to redress this.'

'Britain was the workshop of the world.' That was Fred's great phrase and Neil Carney remembers him saying it time and time again. 'I used to say to him on a number of occasions, "We still make things, you know. It's just that we no longer need a crane to lift them. We make things that go on desks now instead of big concrete blocks." And he used to say, "We were the workshop of the world when we made the heavy stuff but we're not now." He had this great passion for preserving that way of life. He didn't want to preserve the fact that folks were living on a pittance in houses with no modern facilities; it wasn't that he wanted to preserve. It was purely the craftsman's side of the job, the way things were designed and built. He was a great believer in a sense of achievement, "I've done that and it looks right." He was very much consumed with the work ethic, that money should be acquired by productivity, that it should only be made by hard labour.'

It's a view that is echoed by Suzanne Fagence-Cooper. 'A lot of people still use the word Victorian as a term of abuse. It's pejorative, we blame the Victorians for so many of society's ills today. The way we live in our cities, the things we've done to our countryside, the way our factory systems work, we seem to think that the Victorians should have been more far-sighted. I think that Fred, by drawing attention to their great achievements, makes us realize that it wasn't all dark satanic mills.'

When Fred was a young lad, many Victorian fashions and habits still lived on in Bolton. The town had a healthy community spirit and people took pride in appearances. In the rows of terraced housing the windows gleamed, front steps shone white – a credit to the town's women-folk who waged a constant war with the soot and grime. The Bolton housewife would edge her front and back steps with a donkey stone and then tell her neighbour, 'I've done me front and back stoned.' The woman's work of keeping the house clean was what Fred had grown up with and there was no part of him that was a modern man, especially in his attitude to women. 'Basically,' he used to say, 'I'm still like a Victorian when they didn't have any of this feministic sort of outlook. Now it's all this equal rights thing and they want half of everything. For the Victorians the lady's place was to be at home, getting the tea ready at teatime.'

With his old-fashioned views it wasn't always easy to live with Fred, especially for a modern woman like Sheila. 'He was Victorian through and through,' she said, 'and he had certain ideas about how he would expect his wife to behave and respond. It seemed to have slipped his mind completely that as he became well-known on television you needed to be more than just a wife. He'd expect you to be able to run the office and type letters and answer the phone. But if his meal wasn't on the table at a certain time, he'd be quite cross. Or he might expect that you should have the house spotless and even bake bread, but of course there's only so many hours in a day that you can do things, so he'd go out and moan to people. If you called him in for his tea, there was always a bit of an issue on that. There'd be people in the garden at five o'clock and so I'd start the evening meal at a quarter to six. I'd ring the bell and tell him when his tea was ready and he'd come ambling over and say, "Steady on, cock, just a minute. There's a bloke out there that's got unbelievable interest in my lifetime's work." So, teatime would be an hour later and I'd be quite angry and he wouldn't know what all the fuss was about. Fred was very much like Isambard Kingdom Brunel in his attitude to home life, I should imagine. According to Fred, Brunel had this great ability as an engineer, but really wasn't anything like a good family man. He'd leave his wife for weeks on end and they wouldn't speak, and I think Fred thought that gave him carte blanche to treat his wife in a similar fashion. He wouldn't go off and leave for weeks on end, but he wouldn't think anything of spending all day in the sheds and talking to other people. There wasn't much time for his wife, and you had to be quite a strong person to deal with that. It didn't worry me a lot, but Fred thought it was OK to put his wife second. He was chauvinistic, but that was his Victorian sensibilities coming through, part of the things that shaped his world, who he was, what he did.'

Fred was a happy man making the *Victorian Heroes* film. It took him to many places that he'd always wanted to visit, like the site of the *Great Eastern* slipway and the top of the Clifton Suspension Bridge. It was a thrill for him to go to Bristol University Library and study Brunel's letters, papers and diaries and to look at the drawings for the *Great Eastern*. It also gave him the opportunity to visit great Victorian houses like Cragside in Northumberland, the home of Sir William Armstrong, the shipbuilder and arms manufacturer.

Fred's love of Victorian architecture and decoration went back to the time he was serving his apprenticeship as a joiner. 'My work,' he said, 'started to take me into Victorian mansions that cotton mill owners had built – much grander houses than I had ever been used to. When I began to see the quality of the workmanship in them with their fine stonemasonry, architraves around the windows and the doors, beautiful ornate skirting boards, general ornamentation and the sheer Victorian splendour, it intrigued me as to how they did it. What great lengths the Victorians went to, just for something that was good to look at. Whether it was structural, engineering or architectural, a lot of it serves no other purpose than to please the eye. That's what I like about it.'

So popular was Fred's approach to history and architecture that the BBC asked us to develop proposals for a new series in which Fred would look at the history of different types of buildings in Britain. It was the time of Simon Schama's *History of Britain* and BBC2 wanted a history of British building. I had reservations. A series like this would have to be very structured and even more tightly scripted than the things we'd already done with Fred. His greatest strengths were his passion and his enthusiasm and the more scripted a programme had to be the more this started to get lost. I wanted to return to a less structured approach with Fred and had developed a proposal for a series in which he would travel round the country on his traction engine, if he could get it finished in time for filming the following summer. On his journey he would visit some of our best castles, cathedrals and country houses as well as steam and industrial heritage sites. It was to be a record of a journey with Fred travelling round at the pace of an earlier age, meeting people and visiting places that were of interest to him.

Fred was very keen on my proposal, not least because it would give him the impetus to get his traction engine completed and out on the road. So, even though there was going to be another series for us to do, his initial reaction was one of disappointment because it wasn't really the series that was top of our list. But he'd enjoyed all the places we'd taken him to for *Magnificent Monuments*, and he soon began to look forward to a similar trip exploring the history of building, even if he couldn't travel around on his tractor. 'Getting about a bit more in my old age than I used to do,' he said, 'and looking at some of these buildings, some of them have impressed me immensely.' His interest in the great houses, palaces, castles and cathedrals that we visited was not in architectural theory but in the practicalities of how things were built and he was happy with the idea of a series in which he would be able to give simple explanations of how medieval castles and cathedrals had been constructed at a time when there were no power tools, no concrete, no steel, no engines and no heavy machinery.

Fred's horizons had extended significantly beyond Bolton and the steam and steeplejacking that were his first loves. He'd become much more aware that Britain, or England as he still insisted on calling it, is full of magnificent examples of the architectural and engineering genius that we have had throughout the ages, and of the construction skills of the armies of workers who turned the visions of the architects and engineers into reality. In *Building of Britain* he would be able to tell the stories of some of these men and how some of their greatest works came into being. It would be history told from the standpoint of the ordinary working man rather than kings, queens and landowners.

Filming was due to begin in the spring of 2001 but there was an unforeseen delay. Fred had started to have severe abdominal pains and he was admitted to hospital for tests around Easter. At the time there didn't appear to be anything seriously wrong and, after a couple of weeks' rest, Fred was ready for the shoot. He'd always been fit so we didn't envisage any problems. But the stomach pains he had were the first symptoms of a much more serious illness, and from this time on Fred's health would never fully recover.

Within a few weeks of the tests, however, Fred was back on the road shooting the new series. 'The *Building of Britain* series opened a new door or two for me,' he said. 'It were really a history lesson talking about people like Hereward the Wake and things like that. It also gave us access to places on the inside of big castles and cathedrals where nobody ever goes. When we got into the passages inside the walls and all that sort of thing, you could actually see how they'd constructed them.'

Given Fred's background as a joiner, he was particularly interested in the timber-framed buildings that we looked at and the places where this sort of work was still being done today. 'One of the nice things we did were to go on a visit to Peter McCurdy's workshop,' Fred said. 'We'd met him before at Shakespeare's Globe, but when we got to his workshop they were reconstructing a cruck-beam roof for a medieval barn at Glastonbury that had been destroyed by fire. When we were there you couldn't help but get the feeling that his entire workforce had tremendous job satisfaction; making all these lovely joints and all done like they did it in the olden days. The only modern thing they had were like a stacker truck for lifting these great beams up instead of twenty men using their own muscle power to give it a shove. When we went to visit his premises it were quite magic.

'Being a joiner, to visit places like Little Moreton Hall and Harvington Hall and see all the woodwork and the joints and the pegs was very interesting for me. At

Harvington we ended up in the hideaways or the priest holes. To get into one of them you had to remove all the shelves out of a book cupboard and then part of a beam swung outwards and you ended up in this unbelievable claustrophobic slot in between two walls. The trouble was, to get it filmed, it wasn't just me in there. What with the cameraman, the sound man and me all in there at one time I thought we were going to have to get the fire brigade to get us out. To make matters worse, it was a very hot day and we were stifled; there was no air in them priest holes.'

Hugh Dixon of the National Trust told us of the impact that Fred's visits had. 'We know from our own properties that when there's been a programme with Fred on it our visitor numbers will be boosted. The fact that he was there and that he often had a very different slant on our properties was very good for us. I think a lot of people would have been quite surprised at the range of his interests. He rather came into public notice as someone who'd demolished chimneys, people knew him as a steeplejack, then suddenly you have him talking about everything from really complex Gothic structures at Ely to timber work at Little Moreton Hall.'

John Yates, inspector of historic buildings for English Heritage, recognized the contribution that Fred made to our knowledge of buildings. 'Fred was very good at looking and understanding. You've only got to see how he approached the technology of the medieval carpenters who put up the octagon at Ely Cathedral and quite clearly he's really in there, getting into their thought processes. He's understanding how they had put those things together and what sort of machines they needed to make it work and how they put a rope here and a rope there, in just the same way as he would understand how an engineer would assemble a steam engine. Fred could put himself in the shoes of the craftsman, of the people actually making this thing. Most historians would put themselves in the shoes of the monks who had commissioned it, or the rich people who were paying for it. He was actually there in the shoes of the person putting it up, up there on the scaffold and wondering, is this damn thing going to fit? This way of looking at a medieval structure was very refreshing. He's saying how do they put these particular timbers together? What order was it done in? How are the loads transmitted? How did they get those pegs in there? How are they going to maintain this thing? A practical view, and we don't often hear that about old things.'

To illustrate the complexities of the lantern at Ely and how it had been constructed, Fred made a magnificent model. After the programme had been shown on television he was working on his traction engine with a group of his mates. Alf Molyneux was one of them and he recalled how they got their own personal demonstration of the building of the lantern. 'Fred had a wooden model in the yard, so when we asked questions about this after seeing it on TV, he'd have us in front of the model and he'd be pulling bits of string and showing us how they got timbers into position. It made it all very interesting working with him.'

One of the filming trips Fred had particularly fond memories of was the time we went to Glamis Castle. Fred was a great royalist and found it exciting to visit the home of the late Queen Mother, where the Queen had spent so much of her childhood. While we were there, Fred established his own royal connections. 'When the crew were fixing all the tackle up, a rather elegant lady approached me and said, "I'm the Dowager Duchess," you know, sort of related to the Royal Family. I must say that took me aback a little bit. While we were talking she got on about Prince Charles being like me. She said, "I'll get him to give you a ring." But he never did. Even so I enjoyed it up there at Glamis. It were rather good because it's got lots of beautiful turrets that are sort of hanging out in space and when we were filming I was able to get up on the roof and do a bit of steeplejacking to get a closer look at them.'

One of the highlights for Fred was going to the top of Big Ben. We all wanted to do it as a real steeplejacking job with Fred climbing out on to the clock face. It would have been a highlight of his steeplejacking career, but even with all Fred's experience we couldn't get the necessary permissions. There were all sorts of Health and Safety and security issues. Instead Fred had to be content with going up to see Big Ben itself, the great bell inside the tower, and looking at other bits of the Palace of Westminster. He remembered the visit with great pleasure. 'One of the nicest things were a trip to Westminster and climbing Big Ben and then looking at Pugin's beautiful architectural designs and his tiles and his door hinges and his door locks and all of that. I'd only ever been there once before, believe it or not, and I was on my steamroller that time. It was a sort of a protest to lobby Parliament about the road tax on steamrollers. But this time I actually got inside the place.'

But of all the programmes in the series, Fred's favourite was the one on the building of the canals. 'I've always had a great soft spot for canals and to make a film about canals was quite exciting. I'm the chairman of the Bolton–Bury–Manchester Canal Society and, as a kid, I were more or less brought up on the banks of this canal and even made boats and sailed on it as a young man.'

Sheila would go for walks along the towpath with him. 'We used to have a nice walk we went on in summer near Ringley Locks. We used to call them our locks and we'd go up there and we'd go for a nice pub lunch and have a walk round the canal and reminisce about the olden days. It was quite a romantic thing to do really. We did that many times, it was a special place. In fact, there was a house just by there that came up for sale and, in our more romantic moments, we would think about buying it. It was a lovely old-fashioned Victorian house with a lot of land just by the side of the canal. I thought I could just see Fred there; it would have been a showcase for his interests and his lifetime's work. It would have been lovely but I think Fred's heart was so much in the place in Bolton where he'd lived for thirty-five years that he didn't want to move.'

But for Fred there was also an element of sadness about doing the programme. 'It took me back to some of the places that I worked when I were a young steeplejack. When we went to do our filming, the weaving sheds and the cotton mills that followed the Leeds–Liverpool Canal on the whole length of its journey were all derelict. None of the engine houses had any engines in them; the windows were all broken and they had willow herbs growing out of walls.'

The thing that interested Fred most about the canals was the way they had been built and he was very good at showing some of the things about canal building that are difficult to demonstrate, as canal historian Mike Clarke explains. 'There's always a problem with canals in that today most people only see them when they're full of water. Going back and looking at the construction as Fred did was certainly a big bonus. It was much easier for people to appreciate the construction techniques and how you know how deep they are and how much water has to be in the canal.'

Fred was fascinated by all the great engineering feats that were involved. 'The men who built the canals,' he used to say, 'were there before the railwaymen and before the motorway men. Really they were the first great civil engineers in the whole of England. When we were doing that film I got the opportunity to see some of the men who were actually building a canal today.'

David Baldachino, operations manager for British Waterways, took Fred to that canal-building site. 'I can remember taking Fred to a lock on a new navigation that was under construction. It was a very wet February morning and if you can imagine a muddy field where there's a big hole being dug and big pieces of steel being driven into the ground that's just what it was like. Fred was really keen to get to the bottom of this hole and it was quite entertaining to watch the film crew struggle in the mud and the rain with all the cameras and the rest of the equipment to get down in there with him. Fred himself was down there as quick as you like and he spent much longer than they wanted in the bottom of this hole. It was a memorable day for everybody who was working on the site because Fred had spent time with them and talked to them all.'

Fred was recognized wherever he went and whether he was working on his engine in the garden, filming with me or out for a meal with Sheila, he always found the time for anybody who came to talk to him. It wasn't always easy for those who were with him, though, as Sheila began to find out. 'When I first met Fred in 1977, it wasn't too much of a problem because, although he'd been on television and he was well known, he hadn't reached as many people as he did later on. In the early days we'd be in a pub somewhere and you may have got a couple of people coming across to say hello, and that was that. But later on, maybe about 2000 or 2001, we could be sitting in the same pub and it became almost impossible to have your meal in peace. I think at times people would get short shrift from me, but as far as Fred was concerned everybody was welcome. They could have sat down at our table and joined in the meal; for Fred it wasn't a problem. But it was for me.

'There was one instance when we were sitting down in a place in Bolton for our tea on a Saturday night, and of course as soon as you walk in everybody's shouting, "Hiya, Fred." Then in the middle of our meal a man came over with a video camera and started filming us eating. So I shot up out of the chair and told this man not to carry on filming. But Fred was quite happy with it and he just said, "Ignore her, she's bloody nuts. You're all right, mate, you carry on." I was quite angry and cross at that, and mostly at Fred rather than the man.'

The early signs were beginning to show of the pressures that came from being with Fred. And much greater pressures were soon to follow. Fred's health was deteriorating. In spite of the operation he'd had before filming started, his problems with his bladder had continued and were clearly causing him a lot of discomfort, especially when he had to do things like squeezing into the priest's hole at Harvington Hall. Fred had always been tough and his fitness had never been in question. But climbing to the top of towers and clambering around in the rafters of cathedrals for this series had involved a lot of physical exertion and Fred was beginning to feel it. 'In the *Building of Britain* series,' he said later, 'these castles with all their spiral staircases and great inclined planes and all that were quite hard work. It's something I found out years ago. Staircases in big mill towers were always hard work, especially when you had a fifty-sixer in each hand. But now that I'm an old man, even ordinary spiral staircases in the towers of castles are hard work. We went to Harlech Castle and I'd to climb up this great inclined road that came up from the sea and when I got to the top I were really duffed. When people saw it they all thought I were acting, but I weren't. I couldn't get my breath.' The reasons for this were soon to become clear.

11

Back-street Mechanic

Although Fred loved travelling round the country to tell us about Britain's great industrial and architectural heritage, he was always glad to get back to his first love – his own back garden. And there is nothing that sums up what Fred Dibnah was all about better than his garden, which he'd transformed into a one-man industrial heritage centre. Under the twenty-five-foot chimney and the pithead gear that he'd built, rusting boilers and bits of old machinery lay scattered around a somewhat higgledy-piggledy series of sheds and work-shops made from anything that Fred had been able to lay his hands on. Clinging to the edge of the valley of the river Croal, it looked at first sight like a scrap yard, but this place was Fred's idea of heaven. There was nowhere else in the world he'd rather be than here.

From the time he'd first moved into the house, this garden had always been Fred's pride and joy. It was full of old machinery he had salvaged and lovingly restored and everywhere you looked there was another grubby heap of rusting junk waiting for Fred to turn it back into a gleaming machine that once powered the Industrial Revolution. The boiler he had in the garden made steam for turning a steam engine round and that engine powered no fewer than fifteen different machines. While other people tend their flowers, Fred used to stoke the boiler that powered his collection of steam-driven machinery.

Sheila Dibnah recalled the first time she set eyes on it. 'It was a love affair with Fred. I recall when I first came here, Fred showed me all round the garden and explained all about the machines. He was like a proud father telling you things about a child; what each little item of machinery consisted of, what it did, how it had been made. In Fred's mind they were all like family, and he took me round introducing each piece of machinery as if it were an orphan he'd rescued from some heartless mill owner. A lot of it went over my head because it was technical. I couldn't understand what he was saying properly, but the passion was there and I think most people tapped into that passion.'

Fred himself said his great interest in the mechanics of the past had come from when he was quite a small boy and he used to go along the canal on his bike and look at the remains of abandoned engine houses and the like. 'I don't think that has ever left me,' he said. 'That's why all this has come about here in my back garden. Somehow in some strange sort of way I wanted to hang on to the past.'

His friend Alan Crompton recalls the early days when

Fred first moved into the house. 'I remember going down there and the first shed he ever had were only 8 x 4. I think it were even smaller than the one in his mother's back yard in Alfred Street. At that time it was all garden. It were quite nice really; no sheds, no steam engines, just grass. He put his bike in this first shed. And then he wanted a bigger shed. So he got a 24 x 12 end cabin that were badly rotted all round the bottom. So he cut a foot off the base all the way round, dropped it a foot and erected it. The first thing he put in was a little hand-powered drilling machine. Then, of course, he put all his carpentry tools that he'd had from being a lad in there. And he just went from strength to strength then with all these machines. It took a long time mainly because I don't think he bought any of them. They were all scrap – gifts from people who were throwing them out. But with these things he could produce anything. Because they were always archaic and unwanted, they were all belt-driven. But that were ideal for Fred.'

As Alan McEwen recalls, Fred had his own way of doing things. 'One of his favourite pastimes was to journey down to the local scrap yard where you'd see him rooting round all sorts of boxes and skips, searching out little bits of brass and copper. I went with him once or twice and the people who owned the scrap yard obviously thought a lot of him, because they never charged him, or if they did it was coppers.'

It was because of the way he had built it up over so many years that the garden meant so much to Fred. 'I think the garden outside was Fred in his essence,' said Sheila, 'because those things had been gleaned over many years. They were just basically bits of scrap and then they came into Fred's capable hands and he brought them to life again. He turned them into something special. Every time we were away from home I know that Fred was very anxious to get back because he always had another project on the go. It was like having a new baby in a way. If anybody came down to the garden and said, "We've got a milling machine and it's going to get smashed up. Can you rescue it?" Fred would come indoors and moan to me and say something like, "Well, I've no room for it, but I'm not having it smashed up."'

His friend and fellow steam engineer Ian Howard of Alton Engineering recalls the way that Fred used to operate. 'If Fred could beg it he would do. That's how he got on and fair play to the man. He rebuilt all his engines and that sort of work is all expenses, so if you're not helped on your way by your friends it would be a struggle. Fred was great like that. If he could help you in any way, even if it was just sharing his knowledge with you, he would. I did have to smile to myself recently when he was talking about how long it had taken him to do his tractor up. "Twenty-seven years," he said, "but twenty-five of them were spent talking about it!"'

Talking about it and watching other people do a job was Fred's way of learning. 'You can learn a lot more from an artist or a craftsman or an engineer if you're working with him than you can picking a book up and reading it, believe me. You've got to read books to get the basic gist of things but you can't really learn to do things or make things just from reading.'

Fred's way was the old apprenticeship way of learning. It's something that Neil Carney appreciated. 'I served an engineering apprenticeship in the old style in Liverpool. That was how you did it; somebody watching you and if you got all of it wrong, they'd put you right. Of course, Fred was a natural for quick learning. As soon as he felt confident to get on with that boiler, he did. They showed him how to drill all the rivets out and he got on with it. The way he went about things was out of date, but that's what appealed to him. That's the way it had been done when Britain was the workshop of the world and therefore it had made Britain great. So as far as Fred was concerned that was the right way to do it.'

Fred also built up a big collection of old engineering books and magazines. There was one set that he particularly enjoyed reading. 'I got hold of some old Victorian issues of the *Engineer* magazine, and when I first got them I couldn't believe it. I used to lie there in bed reading them with my lamp on until I fell unconscious. The adverts in them are very interesting, especially for things like iron pithead gears, coal tubs, steam engines, locomotives and traction engines. Then it's got news of all the latest inventions, the patents and who invented what and the latest developments in coal mining and in the iron and steel industry. When I went on *Desert Island Discs* with Sue Lawley I had to choose my favourite set of books and I chose these volumes. I thought it were a bloody publication from over a hundred years ago but I got this surprising phone call from the editor of the *Engineer* in modern times and he sounded quite posh. He rang, did this guy, and he said, "Thanks for the plug and we will send you collections of the *Engineer* over the last hundred years in bound volume," which was lovely and it's pretty interesting reading.'

Fred had an absolute passion for simple mechanical engineering processes. He was also a perfectionist, as Neil Carney recalled. 'Many a time I've seen him, if something had gone a bit wonky when I was giving him a bit of a lift with the riveting on the tractor, and he'd say, "We'll come back to that a bit later and we'll put another one in." He'd do three or four more perhaps and then he'd come back to look at the wonky one and if it wasn't right, out it would have to come and we'd put another one in. Now, nobody else would have done that because, if you look at a line of rivets along many a professionally built boiler, they're not dead in line unless they've been done with hydraulic equipment and the very latest techniques. But for Fred everything had to be perfect.'

Fred's perfectionism is one of the things that Alan McEwen remembers most. 'He was an absolutely superb man whether working on a lathe or drilling holes through a pin for a bolt. When he tightened all the bolts up, all the bolts had to go in, in the right order. Everything had to be painted, but prior to that, they'd be greased, and not tightened down, so all the flats on the nut were equal at the front. All the split pins had to face properly.'

Fred was awarded two honorary engineering doctorates for his skills and his knowledge of old-style engineering but in his own mind he was always clear what he was and what the awards had been for. 'I'm a mechanic, a back-street mechanic. The Robert Gordon University at Aberdeen gave me an honorary degree for back-street mechanicking and now Birmingham University have given me a degree for back-street mechanicking so that'll do for me.'

Professor Graham Davies of Birmingham University didn't quite put it like that. 'Fred was given a doctorate,' he said, 'because of his enthusiasm and his promotion of engineering and the history of engineering. He was felt to be one of the foremost enthusiasts for the history of engineering and what it was all about, so it was for his promotion of public understanding of engineering that he was nominated. He came to receive his doctorate in his best dungarees, minus cap but with a mortar board instead. By tradition honorary graduates are asked to give a small speech. For his, Fred didn't go to the lectern. He stood right in front of the audience and delivered a speech right from the heart. It lasted for at least ten minutes and he kept the audience spellbound with his experiences and his history of engineering. He told the students that he wasn't an educated person and that he'd come into television and his history of engineering by accident really and that everything had grown from there. But he was doing something that he loved and having the enthusiasm to carry it through; and that was his philosophy for life.'

John McDougall, a past president of the Institute of Mechanical Engineers, thought Fred's assessment of himself was quite apt. 'I think we are all probably back-street mechanics in our own way. I know very few engineers that don't have an interest in mechanical things and perhaps tinkering with them; it's all part of engineering. It's a practical subject and we ought to celebrate that. Fred was a very good mechanic and an excellent engineer. If you run a 1920s steam engine you can't take it to the local garage to be fixed by computer. You have to be prepared to make the parts yourself, service it, keep it running, and that all has to be part of the fun of owning such a vehicle. So Fred had to be a very good mechanic.

'He is well regarded in the world of engineering because he has this natural enthusiasm for the tasks he does and he owns wonderful toys and gadgets. I think Fred turned engineering into a fun profession. He was very proud of his skills, the machines that he could run, the parts that he could make and building his own steam engines. He also had a very high regard for craftsmen and women. These skills are somewhat undervalued in our society today and Fred tried to bring them to the fore. Universities chose to honour Fred because he's a really good role model. Engineering is often seen as an earnest profession, a worthwhile profession, which is how we market it. Fred marketed the profession as a fun way to work, that would attract boys and girls to the universities.'

But Sheila felt Fred was much too self-deprecating about his engineering skills. 'I don't think back-street mechanic was a good description of Fred really. I used to tell him off for saying that, because it sounded like a disparaging remark in relation to all he'd achieved. I think he was an intuitive engineer. He was brilliantly gifted, he could turn his hands to most things, but when he said back-street mechanic I always conjured up in my mind somebody who was messing around on an old Ford Anglia, so I don't think that was a good description of Fred. He was worth far more than that.'

Fred didn't know the modern way; the old-fashioned way was all he knew and all he wanted. But it did give him some problems with modern life, as Sheila explained. 'I don't think he could cope with the modern world. I find it a great paradox that he could build a steam engine from start to finish, and by contrast he couldn't work a microwave oven or a video recorder. It was almost like a mental blockage with Fred when it came to things that he considered to be modern. He even had trouble working the television remote control. That's not to say he was stupid in any way. It was just that he didn't want to be bothered with anything modern.'

So when it came to building his engine he didn't have a lot of time for something as newfangled as welding. 'He'd call on welders to do an odd job for him now and again,' Neil Carney remembers, 'but he would always rivet everything himself. The expense was another motivation because it was always cheaper for him to rivet something himself than to get it welded.'

Health and Safety was another bit of the modern world that Fred didn't have too much time for. 'Fred was pretty much like myself and my pal Jack,' said Ian Howard of Alton Engineering. 'We can't be bothered with all this legislation and all these safety codes of practice and "You mustn't do this unless you've done a schedule!" It seems like you can't get up in the morning unless you've written down how you're going to get out of bed!'

Fred had very clear views on the subject. 'It's a worrying time now,' he used to say, 'because we're breeding a race of people who do not appreciate danger. They say rules are made for idiots and the rest of us who know how to do things bloody get on with it, which is very true if you think about it. You do not put your hand in between a wheel that's going round fast or shove it in the way of a circular saw. Somebody came in here the other day when I was sawing wood on the bandsaw and he said to me, "Oh, you don't have a guard on there. My uncle used one of them without a guard and he ended up losing his arm." Well, my advice to his bloody uncle would be that he should have changed his occupation to being a gardener or something like that. You don't play with dangerous machinery if you don't appreciate that it's going to kill you if you do the wrong thing. I can't say that I'm unblemished. I've got a few cuts and bruises on me that have been done by machinery, where I did the wrong thing. Nearly always because of some distraction.

That's the thing that causes accidents, being distracted.'

Fred's total disregard for modern Health and Safety regulations meant that his garden could be a dangerous place for filming, because it always involved a lot of movement round his unguarded machinery. With thick leather belts connecting his line shafting to the machines, unguarded circular saws whirling round on their benches, red-hot furnaces for his rivets, an open mineshaft and hot steam all over the place, one false move could have meant disaster. We did a lot of filming when Fred had got steam up and all the machinery in his workshop was in operation. Fortunately we had the same crew most of the time, with Andrew Hartley on camera and Nigel Chatters on sound, so we all knew our way round there and with a bit of care and some healthy respect for the machinery we avoided any accidents. But if we had observed the strict Health and Safety regulations for a film shoot, I would never have been able to allow a crew to step into Fred's garden and some of the best and most informative bits of the programmes would have been lost. But the garden was a superb location and by filming there Fred was able to demonstrate everything from building an arch or a groined roof to the way locomotives would have been riveted in Robert Stephenson's works or how the caulking would have been done on Brunel's *Great Eastern*.

For Paul Jarman, curator of transport at Beamish Open Air Museum, these demonstrations were one of the highlights of the programmes. 'Fred's contribution to our knowledge of industrial and engineering history was to watch him demonstrate in his own workshop at home how you riveted, how you created arches, those practical elements of the Industrial Revolution. It was really good to see things come alive that you've only read about. I remember when he demonstrated caulking on a boiler. I had always wondered what this sort of mystical process was so it was great to see somebody doing it, and the fact that he was doing it in his back garden is something that would be the envy of many.'

Whenever we filmed things like that in the garden I was always conscious of the racket that was being made and wondered whether we'd get any complaints from the neighbours. But Fred always seemed oblivious. Perhaps it was because he was deaf – or, more likely, perhaps he was deaf because of the racket that he'd always made. But how hard was it all to live with? 'It was difficult at times,' said Sheila, 'but I really took it on board because, when I first came here, I could see how passionate Fred was about his engines and indeed everything in the garden, and really that was who Fred was. He was his engines and workshops, so I never saw it as a conflict. I never saw this as being in competition with me and my place in Fred's life. One problem, though, was that Fred was a total perfectionist. This was detrimental to home life at times, because you'd start saying to Fred, "What about the kitchen? Are we going to have a new kitchen or are you going to do this, that or the other?" And he'd start off, but he was so much of a perfectionist that he'd not finish the job because it was never quite up to his expected standards. That and the fact that he never had time to do anything because he was always busy playing in the garden.'

Fred was well aware that his garden was unique. It was one of the few places where you could see things working, and for this reason he always welcomed visitors. Nobody who turned up to see the garden was ever turned away. There would always be many interruptions. As we were working with Fred, a car would park just outside the gates and people would get out and start peering through the railings tourist-fashion. There was a notice in the window saying 'No casual callers', but Fred always looked as though he was hoping that any visitors he had would come down into the garden. If they were not bold enough to come through the gates uninvited, Fred would usually call to them, 'It's all right if you want to come and have a look round.' Of course, once they were in they would usually stop for a couple of hours. 'Plays havoc with my work, it does, but what can you do?' Fred would say. 'You've got to be nice to people. You can't turn them away if they've come a long way to see you. Mind you, Sheila's not that keen.' And that was true. While Fred was always happy to talk to anyone at any time and at any length, Sheila was rather more protective of their privacy.

'Fred is so approachable, he'll let anyone come in,' Sheila said. 'The trouble is we get twenty phone calls a day from people just wanting to come here for a chat.' But the visitors weren't the only problem for Sheila or either of Fred's previous wives. 'The long hours and loving care he lavished on his machines could make it quite difficult at times. Fred always had some project going, and machinery would be going in the background all day. I always had a problem with that because I'd be stuck in the office trying to work and make telephone calls and Fred was just like a small boy who'd decided to play with his machines. It wasn't that I couldn't get out and enjoy myself and do my own thing, but then I'd come back and the machines were still wailing around out there and they'd go on till at least six or seven at night, and sometimes if he was doing a job he particularly wanted to finish they could go on till as late as nine o'clock. So from a wife's point of view it was very difficult to live with. It was almost like living on a works site.

'When I first came here, Fred had two or three helpers each day, but later on, as more work was done on the engine, he assimilated a band of helpers who were only too willing to come down every day. I used to refer to them as "Dad's Army". They were all men of a certain age who had caps on like Fred. They turned out to be great friends to Fred and they helped him a lot, but it was difficult from a wife's point of view, because you were always the odd one out. You were the one who was complaining if they were making too much noise, like a mother figure in a way with small boys outside wanting to play with their toys. They just wanted to carry on regardless.'

Things were now going very well for Fred. His wish had always been to spend three weeks or so earning some money by doing a chimney, then have a few weeks off to play in the garden. Now at last he had enough money to do this. And by this time another source of income had developed. He'd always earned money from after-dinner speaking but he was getting a little bit fed up with it. 'I get invited to speak at some company dinners,' he used to say, 'because the managing director likes good old Fred Dribble. But a lot of the young salespeople and the girls from the office aren't interested in me or what I've got to say. They want to be dancing or chatting each other up, so they all start talking at the back. What makes it worse is that it's usually about eleven o'clock by the time I get up. By then they've been giving me drinks all night and everybody else has had too much to drink so I have to put up with a lot of interruptions.'

But there was an alternative. For some time Sheila and his agent had been suggesting to Fred that he should do theatre bookings. At first Fred was reluctant, because the personal contact with the people he was talking to was important. He didn't like the idea of standing on a stage with lights on him, talking to people he couldn't see in a darkened theatre. But he'd met Billy Connolly, who'd told him it was much easier in a theatre than seeing people in front of you who could put you off. So Fred decided to give it a go. Once he'd done the first few he never looked back. Even though he couldn't see the audience they were attentive, they laughed at his jokes and listened to his stories. Fred realized straight away that his theatre audiences, unlike some of his after-dinner crowds, were all there because they wanted to see him and they'd all paid to listen to him so they didn't start chatting among themselves during his show.

Fred was now better off than he'd ever been in his life. As well as his television money and royalties from books and DVDs, he now had income from his theatre tours and from a number of product endorsements. He wasn't rich by any means, but he was comfortable and could afford to do the things he wanted to do. And Fred still didn't need much money. His old ways of doing things had not changed. He'd still go round the scrap yards for parts he needed for his engines or go to a demolition site if he wanted some doors or windows for a new shed, and he had a regular supply of free wood to provide the fuel for his boiler. Although he enjoyed a bottle of red wine now as well as his Guinness, he'd not acquired any expensive tastes and he was still the last person to dip his hand into his pocket when it came to buying a round — although that was usually because he was too busy talking.

Fred was in the sort of position he'd always wanted to be in. He was completely independent and he would have plenty of time between filming commitments and theatre bookings to finish the tractor that he'd been working on for so long and start on his grand plan to have a fully working replica coal mine in his back garden. Life was looking good. The only cloud on the horizon was his health. In the summer, while he'd been filming *Building of Britain*, his abdominal pains had returned and begun to cause him a lot of discomfort, although he never complained about it.

Towards the end of filming he went back to Bolton Royal Infirmary for a check-up and a tumour was found in his right kidney. When the kidney was removed, the tumour was found to be malignant, so Fred was sent immediately to see a cancer specialist at Christie's Hospital in Manchester, where he had a scan. The results were devastating. The cancer had spread and cells were found in the fatty tissue that had surrounded the kidney. But that wasn't all: the cancer cells had also travelled down into his bladder. Fred was told he'd got about a year to live if it wasn't treated immediately. The hospital could offer chemotherapy treatment that might give him a bit longer, but basically it didn't look good.

Within an hour of getting back home, there was another major shock. Sheila's father was working on the roof of his house that afternoon when he slipped and fell to his death. Looking back on that traumatic time, Sheila feels that her relationship with Fred began to deteriorate from that point onwards. Fred had been at a low ebb before, when each of his wives had left, but there had never been anything like this. More than anything else he felt a sense of anger and injustice. He'd worked hard all his life and now, just as things were getting comfortable and he could afford to do what he wanted to do, his life was being taken from him. 'It isn't fair,' he would say. 'What have I done to deserve this?' He'd got so much left to do and so little time left to do it. After nearly twenty-five years of working on the engine, it didn't look as though he would ever see the day when it steamed out of the drive and on to the road. For Sheila it was equally difficult. She had her own grief to bear and she and Fred found they couldn't help each other very much. Fred would get increasingly hostile and, in his darker moments, would say, 'What have you ever done with your life? It's me who is going and you're not. But what have you got to offer?'

Fred went for his chemotherapy treatment. He hated being at the hospital with all the waiting around that it involved, but he soon found there were interesting things to occupy him there. He made friends with the boiler man and would spend his time in the boiler room talking to him and to some of the hospital maintenance men. It kept his mind off what he was really there for. He'd also make friends with other patients and would invite people home who wanted to see his garden. Alf Molyneux, who was there with Fred every day helping him with the traction engine, remembers one of these visits. 'This chap came in, another patient he'd met at the hospital, and he was stood at the side of the road swapping medical notes with Fred. Fred said to him, "Have you got your date yet?" So this fellow said, "A date for what?" Fred replied, "A date for dying." I tell you I started laughing, I couldn't help it, and Fred started grinning and said, "What are you laughing at?" I said, "Do you know what you've just said?" I repeated what it was they'd said and all three of us started laughing about it. He'd said it so matter-of-fact.'

Fred wasn't going to be beaten by the cancer nor was he going to be put down by the chemotherapy treatment he was receiving. Nobody who saw Fred at the time could believe the way he came back after the treatment and apparently shrugged off the debilitating side effects that chemotherapy generally has on people. Against all medical advice he would get back home and that same night he'd have a couple of cans of Guinness and a few glasses of red wine. Fred was as tough and stubborn as he'd ever been and he wasn't going to give in easily. Within a day or so of the treatment he'd be back out in the garden getting on with the work on the tractor and everything else that time was running out for. And that toughness saw him through. When his course of treatment finished he had to go for a scan and the results were positive. The chemotherapy treatment had been successful and Fred was given the all-clear. Life could get back to normal.

12

Age of Steam

Once he'd got the all-clear Fred worked on the engine with renewed vigour. He knew now that he was going to get it completed and there was an added bonus: he'd got another chimney-felling job. This one was going to be a bit of an occasion. It was on the site of an old mill at Tottington, on the outskirts of Bury. Kirklees Mill had been demolished to make way for a new housing estate, leaving just the chimney standing. Fred got together the old team of Eddy Chatwood and Mick Berry and they spent the best part of a week doing all the usual preparatory work. The actual felling was to be at lunchtime on a Sunday and it would make a major PR event for the demolition company and the housing developers. The mill had been in a little valley and the flat high ground just above it provided a perfect vantage point. A big marquee was erected there and council officials, planning officers, land agents and prospective customers were all invited to the spectacle. On the morning of the drop Fred and his team were there early to finish their preparations. As they built up the fire, wine and beer were delivered to the marquee and the caterers arrived with their sandwiches and finger snacks.

The felling was scheduled for one o'clock and from around twelve the invited guests began to arrive. Drinks were served and glamorous-looking PR girls milled around. It was all very different to Fred's early chimneys. Some of the guests started to secure good vantage points on the viewing platform that had been erected outside the marquee, but the weather began to turn. A cold wind blew up and within minutes it was raining. Drinks in hand, a few hardy souls kept their viewing positions but most drifted back into the marquee as Fred and his team struggled in the mud at the base of the chimney. Then, just as it was time to light the fire, the weather got really nasty. The wind turned into a gale and the light rain became a downpour, hitting the chimney horizontally as it swept down from the nearby moors. Everybody retreated to the food and drinks inside the marquee. Only Fred and his men were left to brave the elements. But there was a decision to make. The wind and the rain were blowing in the wrong direction. They were so strong there was a chance they could affect the line of the fall. The team stood around in the pouring rain waiting to see if conditions improved before lighting the fire. It was a big clear site and a lot of people had turned up to watch the chimney come down, but better to be safe than sorry.

The rain didn't let up, but after half an hour or so the wind began to drop a little. Fred decided it was safe to go ahead.

The wood and the tyres for the fire were soaking, so lighting it was difficult and took a lot longer than it would have done normally. Eventually, as the rain continued to pelt down, the fire blazed and the chimney dropped in exactly the right spot about an hour after its scheduled time. A few of the guests had made their way to the edge of the marquee just in time to see it come down but the majority were warm and dry and on their third or fourth glass of wine and well into demolishing the prawn sandwiches. Fortunately for Fred, there was still some Guinness left as he came into the tent soaked to the skin but triumphant.

The next day he got a phone call. It was from a lady who had a house that overlooked the demolition site. She asked Fred if he'd got another chimney-felling scheduled because her young son had been looking forward to seeing this one. He'd sat at his bedroom window all morning watching the preparations, but just after one o'clock he'd been called away to have his lunch. When he got back to the window the chimney had gone. For the rest of the day he kept saying, 'I missed it. I didn't see it come down.' Fred thought back to the time more than sixty years earlier when he'd sat with his classmates on the field at the back of school waiting to see his first chimney-felling and his disappointment when he got back after lunch to find that it had fallen in his absence.

Soon after this I went over to see Fred about doing another series. We'd had a good run so far and Fred had always enjoyed the filming. He was keen to get some more of these holidays paid for by the BBC, as he still regarded them. Up to this time we'd never done a series on steam and I knew it would be a real treat for him if we could get one commissioned. He would be able to indulge his greatest passion and for Fred that would be the pinnacle of his television career. We talked about steam and how that passion had run all through his life.

'Yeah,' he said, 'a steam engine really is a fascinating thing. When it's actually running it comes alive in a strange way. It has an unbelievable smell for a start. There were an old guy came in to see my boiler the other day, eighty-odd years old he was. He was sniffing away and he said, "That brings back memories from me youth, that smell of oil and steam is a smell all of its own." It has been said if you could put it in a bottle and cork it up you could sell it, it smells that good. And then there's the noise that they made. Some of them were very quiet, but if you had a great pit winding engine, winding in a shaft that were a thousand yards deep with a cage hanging on the end of the rope with maybe ten tons of coal in it, the engine would make a terrific noise when it got off to a standing start. There were a winding engine at a pit called Parsonage and you could hear it in the next town down the road when it set off. The great cloud of steam over the top of the engine house was a sight that the modern man has missed. Even now the nearest you can get to it is a

steam locomotive going up a big hill with a lot of
carriages on and everybody goes into raptures. But the pit
winding engine were something else; the roar were like
twenty locomotives going up a big hill with a lot of coal
behind them.

'They were wonderful things, steam engines, and it's
amazing, isn't it, when it totally relies on a man to keep
it going. If it comes to any big hills tons of coals have got
to be shovelled because it gobbles it up. Even my steamroller in its humble way needs a shovel of coal every
other lamp post when it comes to a hill and by the time
you get to the top it's gone.

'Well into the twentieth century steam locomotives
were still being built. They operated on the railways right
up to the 1960s and lots and lots of big mills in Lancashire
were still driven by great big reciprocating steam engines
until this time. Sadly all these things have gone now, but
steam is still alive and well in my back garden. The boiler
I've got makes steam for turning a steam engine round
and that engine works no less than fifteen machines. It
comes in handy for lots of occasions and particularly for
repairs to the traction engine. The tractor is now almost a
kit of parts that's ready for assembly – just a little bit
more finance and time and I could complete it. It will be
an unbelievable sense of achievement when I get it
finished because by then I would have built the whole
thing myself from scratch with a complete new boiler and
all the parts hand-made. I've made new piston rods
myself and new covers and valve rods for the valve chest

covers and taken off all the corroded parts to get them back looking like new and even better than new, some of them.

'The vertical boiler I've got out in the back is very primitive. It's the simplest form of boiler next door to a kettle. Basically it's just a cylinder full of water with a very shallow firebox at the bottom of it and a tube straight up it for the gas and the smoke to escape through the atmosphere and to get a bit more heat out of it. The crudest ones like this one just have two tubes going from one side of the firebox to the other; all the water is in those two tubes and the flames have got to circumnavigate round them.

'When they were being used commercially coal was dead cheap and these vertical boilers would burn any old rubbish – wet coal dust they would fire them on, stuff they would normally throw away. It's amazing really how late on they survived on British Rail. All the breakdown trains were steam-driven and all had vertical cross-tubed boilers. The horizontal boiler I've got is quite a modern one and the design is good because the circulation of the water in it is very good and that's the most important thing with a steam boiler.'

Steam was a massive influence on Fred's life and he was very knowledgeable about it. John McDougall of the Institute of Mechanical Engineers summed it up. 'I think the biggest influence on Fred was his love of steam and his interest in how it can be used to power large-scale machines. It would have been easy for him to have powered his workshop with electricity but he chose to build a great big enormous boiler, great big enormous flywheels and drive his machines from that. It probably took him something like three hours to fire up his workshop, which he could have done with a flick of a switch, so I think steam was the big driver in his life.'

Because of this passion there was no doubt in my mind that, if there was to be one more series with Fred, it would have to be one on steam. In it Fred would share his enthusiasm as he unearthed the best of Britain's steam heritage and went out to meet friends and fellow enthusiasts who devote their lives to finding, restoring, preserving and operating steam locomotives, traction engines and stationary engines, mill workings, pumps and pit winding gear. There had never been a history of steam power on television, so as he travelled round Fred would tell us about the part it played in the Industrial Revolution and show us the machinery that made that history. Throughout the series Fred's approach would be very hands-on: he would live every schoolboy's dream, travelling on the footplate and driving the engines.

Fred Dibnah's Age of Steam was commissioned by the BBC in September 2002. Again there was to be a book to accompany the series. We did some preliminary filming at Fred's house in November and he talked first of all about how his interest in steam had developed when he was a small boy. 'It's a strange thing really,' he said. 'As a little lad I was surrounded with great tall chimneys that fascinated me. They all had great clouds of smoke coming out of the top a bit like a Lowry painting and of course when you wandered up various back streets you could hear the rumbling inside these great spinning mills. We lived in close proximity to the engine sheds, where on me way home from school there'd be literally dozens of locomotives lined up all steaming and hissing with water dripping out of them. I used to sneak in and have a look round. Then later on, when I became a joiner, I used to get really brave and sneak into mill yards. I'd be full of fear, climbing up the steps of engine houses and looking at the thing going round through the window and seeing the engine minder in an easy chair snoozing. But he wouldn't really be asleep; he'd be listening for any change in the pattern of noise that were coming from the thing.

'To be an engine driver, or an engine "tenter" as they were referred to in Lancashire, were a very responsible business. When you looked through the window the guy sat there in an easy chair might have looked as though he were half-asleep but he'd be listening for any slight deviation in the noise that were drumming on all day in his head. If it just made a few little funny noises he'd got to find out what that were because they couldn't afford stopping it. If the engine stopped it were like disaster both ways; not just for the management but for the workers as well because, whenever the engine stopped, they weren't earning any money. Nobody liked the engine tenter if the engine stopped.

'They were very impressive; to see a big mill engine

with a thirty-five-foot flywheel connected to wheels and line shafting up five storeys of a spinning mill all going round and almost silently was an amazing sight. They were incredible pieces of machinery and now it's so sad because you go round places like Oldham and Rochdale and all that's left are these gaunt empty engine houses.'

We started filming *Fred Dibnah's Age of Steam* in February 2003 and it was appropriate that one of the first places we went to was one of these great engine houses that Fred remembered so well. 'Trencherfield Mill, at the place called Wigan Pier that George Orwell made famous,' said Fred, 'is a very good example of the sort of engine house I used to see in Bolton. It's got a very good preserved example of a John and Edward Woods engine made in Bolton. Somehow or other they managed to transport it all the way to Wigan and erect it here.' With its enormous twenty-seven-foot flywheel, the four-cylinder, triple-expansion engine is one of the world's last working mill steam engines and it is still connected to the rope drive system that took the power to all the floors of the mill.

But Wigan Pier brought back more recent memories for Fred. Soon after he'd married Sheila, he'd taken her there on his idea of a holiday with her son Nathan and his two lads, Jack and Roger. They'd all gone there with the steamroller and the living van in tow and parked up in the car park at Wigan Pier for the night. It was the first and last time that Sheila sampled the delights of staying in the living van and she was less than impressed by its lack of toilet and washing facilities, especially with five of them in the rather cramped space. But that wasn't all, as Fred remembered. 'We'd got settled down in bed eventually when we heard voices outside. It was a young couple on their way back from the pub. When they saw the living van they got up on to the running board and got carried away with the romance of it all. The whole bloody living van started to shake with them having it off outside and us with three kids inside.'

Our first night away on the *Age of Steam* filming trip was no less eventful. We were staying at Stone in Staffordshire where we were going to film Fred with David Hulse, a very skilled model maker who had built some magnificent scale models of early Newcomen and Boulton & Watt engines. Our hotel was in the centre of town and our rooms were in an annexe separated from the main building. After having a meal at the hotel we set off in our customary manner to look for a local hostelry. It didn't take long. We found a very lively one just over the road from the hotel and as soon as we walked in I knew we were set for another lock-in. They'd become an occupational hazard of filming with Fred. This time was no different. A crowd gathered round Fred and the pints were lined up on the bar as Fred launched into his stories. By 11.15 Fred was in full flow and the audience were hanging on to his every word. More pints appeared on the bar and the door was locked. There would be no escape for another few hours. Wherever we filmed, the people in the pubs we went into didn't appear to have any jobs to get up for the next morning. Some people would have stayed all night listening to Fred's stories and, as the pints went down, Fred always looked as though he would be happy to oblige. Round about one o'clock I would begin the task of trying to prise Fred away from the group. It was never the easiest of jobs and it didn't make me very popular, but eventually Fred would say, 'He's the boss. We've got to be up in the morning, so I'd better go.'

That night in Stone we stepped out of the warm, cheerful lounge to one of the coldest February nights I can remember. We didn't have any coats with us because the pub was only just across the road from our hotel, but when we tried to get into the annexe we were staying in the door was locked and our keys wouldn't open it. Not to worry, we'll try the main building, we thought, but that too was locked and all the lights bar one on an upper floor were out. We threw stones at the window and eventually a head appeared. It was one of the catering staff but he didn't have any keys so he said he'd call the manager, who didn't live far away. As we waited outside for the manager it was freezing; even Fred was feeling the cold. We'd just started to think about sleeping in the car when a young manageress turned up. She tried opening the door with her master key but it wouldn't open. Somebody staying in the annexe must have bolted the door on the inside, she said. The only way to get in would be to knock on the windows of some of the rooms to wake somebody up and get them to open the door for us. By good fortune Jack and Roger were with us that week. It was their half-term holiday so they were over from the Isle of Man and sharing a room with their dad. By trial and error we worked out where Fred's room was and after a bit of hammering on the window a sleepy-looking Jack came to the door and unbolted it.

Our filming got off to a good start, but then there was a delay. Although Fred had been given the all-clear, he was still having problems. Scans revealed that he had several polyps in his bladder so he had to go into hospital again in March to have them removed. He recovered very quickly and within two weeks he was back on the road doing some theatre bookings. By now Fred was doing a lot of theatre dates so he had to fit the filming around them as well as his operations and check-ups. It meant that our next week's filming wasn't until April. It was in Cornwall. We went back to see the Levant engine and managed to get it filmed properly with Fred and one engine minder rather than the whole of the Greasy Gang. If February in Staffordshire had been cold, summer had come early to Cornwall with clear sunny days all the time we were there. We went to Cornish Engines at Pool then I wanted Fred to do a short piece to camera on a beach. He'd talk about how today Cornwall is associated with beaches and holidays in the sun, but two hundred years ago it had been a cradle of the Industrial Revolution and many early developments in steam power had taken place here. We went to Sennen Cove, not far from Land's End. It's one of Cornwall's most beautiful beaches. Family groups played on the sand and windsurfers glided across the bay in the spring sunshine. This really was a bit like being on holiday. But not for Fred. He didn't like it and I've never seen anybody looking as ill at ease on a beach as Fred did that spring afternoon. He summed up his views when he looked out at the windsurfers and said dismissively, 'I can't see much fun in that. It would be better going along the Bolton–Bury Canal on an old oil drum than that.'

For Fred, the highlight of the Cornish trip came at the very end. We were going to attend the Camborne Trevithick Day, to commemorate the engineer and inventor of the high-pressure engine. Fred would be filmed riding on the replica of *Puffing Devil*, which was to lead a parade of traction engines up Camborne Hill where Richard Trevithick had taken the original for its first run. We met up with the engine and its crew on an industrial estate on the edge of town where they were getting steam up. Fred did his scripted pieces about the engine and then got on board to have a go. The idea was that they would drive up and down the roadway inside the works grounds so we could get some shots of the engine with Fred on board. As the engine steamed past the camera with Fred hanging on, he was beaming. The engine trundled down to the factory gates, but then, instead of turning round as it was supposed to do, it disappeared through the gates and headed for town. We'd lost our engine and our presenter. All the roads were closed for the event so we had to follow on foot.

When we got into town the main street was lined with

traction engines. Crowds milled around them and smoke hung in the air but there was no sign of Fred. Eventually we found him at the far end of the street, chatting to an old mate who was there with his engine. A crowd had gathered round; photographs were being taken; pieces of paper and programmes were being thrust at Fred for him to autograph. It didn't look as if we would get much filming done, but we decided to have a go. We followed Fred along the line of engines and he stopped at each one as he was greeted warmly by the owners – long-lost mates from the traction engine world meeting up. Another gentleman with a video camera followed us. Whenever I set up a shot he stood between our camera and Fred and blocked it so that he could get in closer. One of the organizers came along and tried to pull Fred away so that we could get a shot of him eating a Cornish pasty. When I said that was not what I wanted, he wasn't very happy. It was chaos, for all the reasons why I'd never done any filming with Fred at a traction engine rally. Even Fred found it was getting too much, so we decided to move on to the peace and quiet of Trevithick's cottage.

But the episode was typical of Fred. Once he got to a traction engine rally he was very much one of the lads. Andrew Semple, vice-president of the National Traction Engine Trust, says, 'The thing I admired most about him was the way he would go to an event in two roles. He would go to a rally as a publicity attraction and his name would be advertised and he was there to meet the public. But equally he would go to many events just as

Fred Dibnah, engine owner, rallying his own roller. It didn't matter which role he was in, he would always have time for individuals. I've seen him sit at a table for two or three hours at a time, signing all sorts of odd bits of paper in that beautiful copperplate handwriting and always asking who was it for and what was their name and how do you spell it. It always took time and it didn't matter if he sat there till whatever hour of the day, if somebody was waiting to talk to him, he would find time to do it. I admired that. Watching him, you felt that it was tiring and difficult and people were very demanding because they regarded him as public property. But it didn't seem to matter to Fred. All you had to do was to put a pint in his hand and wind him up with just a little memory jogger and off he would go and you could stand back and enjoy a great story. Sometimes when he told a story the next time round it wasn't quite the same, but it didn't matter really because it was good fun.'

'I think he revelled in his celebrity status,' Sheila said. 'I had a word for it, I called it playing to the gallery. I don't think it was because he was egotistical or that he liked the spotlight just for the sake of it. I think it just presented another opportunity for Fred to get his message across, because anybody who was interested in what Fred was on about would be his best friend for life. But he did love being the centre of attention. We would go to interviews sometimes, maybe on a radio station, and somebody would say, "So, Mrs Dibnah, what do you think?" and Fred would say, "Hang on a minute, I've not told you that story about such and such." I could never get a word in!'

Our next filming trip was in May, when we visited the Bluebell Railway in Sussex. Fred enjoyed filming at the preserved steam railways because it always gave him the opportunity to ride on the footplate. The Bluebell Line had four locomotives in steam on the day we visited and Fred travelled down the line on all of them and had a go at driving. 'Really,' he said, 'the programmes about railways were quite wonderful. Plenty of footplate rides on locomotives and lots of visits to museums and nice places where they were renovating steam engines. Unless you've had a ride on a steam locomotive travelling at speed, you have never lived. It's a very exhilarating business. I've done a bit of driving myself. It's a great feeling – the whole thing develops a rock when it's working and you've got to have your sea legs to stay vertical.

'One of the most exciting things I ever did was when I went to do a talk and part of the deal was a ride on an engine. It was a Great Western and fairly old, from about the 1890s. The tender on it was enormous but very low. We were going backwards but you could see straight across the tender. They gave it to me to drive and we were really belting along. I had to shout to them, "I've never been on here before. If we're getting too fast, you'll have to give me a nod." We were shoving the tender of course and we went down this hill and through a tunnel and it were fantastic. I don't think it would have been half as exciting in an engine with an enclosed cab. The fact that you are standing on the footplate with all the wind coming straight in and you are in charge of it all; you just felt if anything got in the way you would never be able to stop it. I must have been bordering on the limit and I've got to admit it was good. It's something everybody should have a try at.'

When we were filming with Fred it was important to make sure he enjoyed everything and not to be too regimented with him or insist that he was word perfect on his pieces to camera. As long as he was having fun, the humour and the natural enthusiasm would bubble over. He was easy to work with, always happy to go along with whatever he was asked to do. However many times he had to say a line to get it right, he always accepted it with good humour. The fact that we had a settled crew and production team was a great help. Fred knew everybody and was very relaxed, and he was brilliant at passing on this laid-back approach to other people we were filming with him. It's not easy being on camera when you are not used to it but Fred would just say, 'Ignore this lot. Just pretend they're not there. That's what I always do.' And it worked. Fred captivated the people he was talking to and they relaxed with him. It had a lot to do with the way he talked. Because he didn't speak in the well-modulated tones of a television presenter, and because he didn't always wait for an answer as most television interviewers would do, a lot of people seemed to forget that the camera was there.

However many times we filmed with him, though, he

never understood the basic mechanics of filming. He seemed to have a mental block about it – a bit like microwaves or television remote controls. From the first day to the last he never really grasped, or at least he pretended never to grasp, why we had to do things two, three or sometimes four times. 'I've already said that,' he would say when we needed to cover something from another angle in order to be able to cut a sequence. 'They might not look as though they know what they are doing,' he used to say to people he was interviewing, 'but it all turns out right in the end.'

Fred would usually have a radio mike on when we were filming. Very often it wouldn't be turned off between takes and he'd wander off to talk to people on the site. Many a time on *Age of Steam*, Nigel Chatters, the sound recordist, would hear him talking about the technical detail of an engine that was being restored or about some of the latest goings-on in the world of steam. Then Fred would look over at us and say, 'They live in a world of their own, that lot, you know.'

Interviewing was always a bit of a problem, especially on *Age of Steam* where Fred already knew most of the answers to the questions. 'Ask him what this engine does,' I would say to Fred, 'then ask him how it works.' The camera would start rolling and I'd cue Fred to start the interview. He'd turn to the interviewee and say, 'Now then, John, we know that this engine was built to pump water from these mines, don't we, but that lot there don't know that, and we know that this is the way it works.' By the time Fred had given a long answer as part of the question, all his interviewee could say at the end was, 'Yes, Fred. That's right.' Fred didn't like anybody to think he didn't know the answer himself.

But this wasn't just something that happened when we were filming. Sheila spotted it many times. 'Sometimes when he got very overenthusiastic about his subject, he'd talk at you rather than with you, because he's got so much information he just wants to download. He just couldn't stop and then he'd go off at a tangent and tell another tale. He was never stuck for words. But the thing about Fred was that where you might get some other guy propping up a bar and telling a tale without knowing what he's talking about, Fred did know what he was talking about because he could do it himself.'

Age of Steam gave Fred the chance to travel all over Britain to see the things that he was passionate about and to pass on his enthusiasm to others. Fred featured on the pages of the steam and vintage magazine *Old Glory* and its editor, Colin Tyson, said, 'Fred was just so happy whenever he was near a steam engine of some sort. This was true whether it was a huge beam engine or a traction engine in a field. That absolutely came out in everything he did, everywhere he went to and in the historic vintage equipment that he used to play with in the yard. Everything about Fred was steam and vintage equipment and it was very good for the whole traction engine and railway movement. He was our greatest mascot. Never was he happier than when he was on the top of a beam engine looking down.' Fred's enthusiasm was contagious.

A lot of the series was filmed in the garden, because Fred had all the steam power we needed on tap and he could demonstrate just about anything when he raised steam there. For Fred it was all playing in the garden. He was for Sheila 'the eternal child. But unlike a child,' she said, 'he had this great knowledge and he was able to apply that and eventually make a living out of his hobby. It was all like wish-fulfilment for him. Fred could go out and play all day on his engines, and then somebody would ring up and say, "Do you want to bring your engine to our rally and we'll pay you x pounds?" That was very good for Fred, because I can't think of anybody else who'd find themselves in a position where they could make a living from their hobby.'

Fred was now making a good living from his hobby and he could afford to do what he wanted. Presenting the *Age of Steam* series had been his dream job. It had taken him to places he loved visiting, and given him the chance to drive all sorts of engines and to meet up with a lot of his old mates. While we were filming, Fred was able to get on with the work on his tractor and use it for some of his demonstrations. 'Twenty-seven years and two divorces' was his favourite phrase and now the project was nearing completion. 'I didn't mind the twenty-seven years bit,' Sheila said, 'it was the two divorces that got on my nerves. Everywhere we went, Fred would say that and he'd say it in a funny way to elicit some kind of response out of the people who were listening, but it always used to make me cringe. I didn't mind the engines so much, because that's what Fred is. He belongs to a world of steam and you've got to accept that's what you married into when you married Fred. But I didn't like the constant references to the two divorces.'

After all the years of hard work, his own traction engine was nearly finished. With all the help he was now getting from Alf Molyneux and Jimmy Crooks and the other members of 'Dad's Army', Fred was confident he would have it on the road by the following spring. But his health was still giving cause for concern. One day he returned to Christie's for a check-up and his worst fears were confirmed. A large tumour had appeared in his bladder.

Fred embarked on another course of chemotherapy. But like the first time, he wasn't going to let it stop him doing the things he wanted to do. The doctors ordered complete rest, but Fred was a man on a mission. He wasn't going to take any notice of advice like that when he'd got his traction engine to finish. And it wasn't just the engine, because Fred had discovered another passion and it was beginning to take up even more of his time than the engine.

13

Dig with Dibnah

Fred was now a man with a very clear mission. He was sixty-five years old and he now knew that his cancer was incurable. The second course of chemotherapy hadn't worked and the cancer was too advanced for an operation. Fred knew that time was rapidly running out, but he was tough, and he was determined to accomplish as much as possible before he became too ill. He had two great ambitions: to complete the restoration of the traction engine and to sink a replica coal mine in his garden while he still felt fit and well enough to do it. He was confident that, over the next six months, he would achieve the first of these ambitions – the building of the Aveling & Porter road locomotive. By now the engine was looking good and he'd reached the most exciting stage of the rebuild. The wheels were due to be put on, the cab attached and all the shiny green cladding and the polished brasswork had to be added.

It would be a great day in Fred's life when his traction engine steamed up his drive and on to the road for the first time. But, great as the challenge had been to build the engine from scratch, it was nothing compared to the other challenge that Fred had taken on. His whole garden was dominated by the forty-foot-high pithead winding gear that he'd erected several years earlier, just after he'd restored the blunger and the steam engine at Wetheriggs. It was the first stage of his most ambitious project yet: to sink a hundred-foot mineshaft, brick-lined all the way down just like a real pre-Thatcher-years working colliery. Some would say it was a pretty madcap idea. People have made some strange things in their back gardens, but nothing quite as eccentric and on such a grand scale as a fully working mineshaft complete with winding engine.

Fred explained the thinking behind it. 'Some years ago I got this hair-brained idea to build a simulated coal mine. It struck me with quite a lot of concern that whenever you went to a preserved colliery museum, it weren't really quite like what it used to be. It's very sad when you go to look at them because they're just a shadow of their former selves. When they were winding tubs up in the olden days, the top of a mineshaft were very violent. There were unbelievable action and a lot of bell-ringing and clanging and what have you, boom, boom, ding, dong, dong. Gone quick as that! What I want to create is a scene like that. What would happen is, we'd have a hundred-foot-deep shaft and a tunnel to the river in the valley down below with an endless rope haulage system.

Mind you, we're not going to be mining coal; we're just digging a bloody hole.'

The first stage in bringing this grand vision to reality had been to build the pithead gear. 'It is really a bit unusual to have a pithead in your back garden,' Fred said, 'especially if you live in a reasonable residential area like the one I'm in. But what a lot of people don't realize is that in Darcy Lever, just down the road from here, there were once about nine collieries and there was a company called the Darcy Lever Coal and Terracotta Company. So there were a lot of pits round here and all I'm doing is recreating one in my own back garden. Most of the people who owned these pits lived in this neighbourhood; this where I live were the posh end. The bit down the road used to be called Dolly Tub City because the women took washing in to subsidize the poor money their husbands got working down the mines.

'I've been interested in mines since I were a kid. The earliest memory I have of coal-mining around here is as an eight-year-old riding my bike along the canal towpath with my dad and I was concentrating on making sure that I didn't ride off into the water! There were lots of interesting things along there, relics of old industry and of the mines, and because it was a bit water-bound you couldn't get near some of these remains with a motor vehicle, so the scrap men hadn't come with their wagons. It meant that, although these places had been shut for quite a few years, things like the pithead gear and the remains of old steam winding engines were still there and

I always promised myself that some day, if ever I got rich enough, I'd build one.

'Ladyshaw Colliery, a bit further up the canal, were quite a splendid place. It were shut in 1949 by the National Coal Board and I never actually saw it working, but when I did get to see it just after it had closed it were summat else. The canal were quite wide and it were full of sunken canal boats. The prows and the sterns were all sticking out of the water and you could see all the rest of the boats under the water. It were lovely and clean because there was no more activity there; no doubt it would have been a murky brown colour when the pits were still going. All the boats had *LMS*, for the London Midland and Scottish Railway, or *Ladyshaw Coal Company, Little Lever* written on the sides. The boat-building shed looked rather like a railway station with beautiful timber boards. It were like a time warp and it all looked so sad really. Every time I was going to see this strange place I got the feeling of anticipation and excitement – it were magic. You could actually stand at the top of the shaft and see the cage hanging on the rope and hear all this water swishing and all these strange noises. After going there about three times, we turned up again, but to our bitter disappointment the wires, wheels, chimney, wagons and railway were all gone. The last time I went there to look, there were just a mound of grey shale where the hole had been and any evidence of its magical past had all disappeared.

'Going back along the canal towards Bolton were the

remains of another pit called Farnworth Bridge Colliery. I remember the ride I did past there with my father and the first time we went you could still see the winding engine painted green and gas lamps and the ropes still on over the wheels and the cages and all the pit tubs and everything all still around. Then maybe twelve months later, which to a little lad seems like for ever, we went there again and the ropes had gone off the wheels and the engine had gone. The winding gear was still there and somebody had knocked a big hole in the brickwork around the shaft and you could actually throw bricks down. When I were a small boy all the mineshafts were still open down there and I used to go hollering bricks down the shafts and when you whizzed a brick down it made an unbelievable noise. Boom boom boom, echoing all the way down.

'There were quite an interesting sprinkling of coal-mining oddities near Burnden Park football ground near where I was born. My aunty used to say that there was a coal mine underneath her house and of course when she popped her clogs and went upstairs, these houses all became empty and they never resold them. They were all demolished and I don't know why but even now there is a fence still round them. It's as if they smelt danger there and the council knew about it, so they never built anything on that spot. It's still like that to this day so my aunty must have been right.'

From those early days Fred always had a great interest in coal-mining. 'I don't think I would ever have liked to

CLOCKWISE FROM TOP: Miners at
Victoria Colliery, Deane, Bolton,
September 1947; preparing to
go down for a shaft inspection
on top of the cage at Big Pit
Mining Museum, South Wales;
the winding engine in Fred's
garden; Fred's pithead gear.

Fred's last chimney – Park Mill,
Royton, near Oldham.

TOP AND CENTRE: Setting off from home to film *Made in Britain*.

BOTTOM: Llanberis Pass.

Weighing the
engine at Crock Hey
opencast mine.

Jack and Jimmy
Crooks lend a hand.

Oiling the engine.

A thoughtful moment.

Running repairs
at Coniston.

Cornish Engines,
Pool.

LEFT: Jack, Fred, David Hall and Alf Molyneux take a break during filming in the Lake District.

BELOW: Independent Forgings, Sheffield.

Life on the road.

ABOVE: Fred chatting to Gordon Newton at the Israel Newton Boiler Works.

LEFT: Peak District National Park in Derbyshire.

TOP: The only way to see the sights in London.

MAIN PICTURE: Fred collects his MBE, 7 July 2004 – a proud day for a royalist.

have been a hewer of coal,' he used to say. 'It's the engines and all the winding gear and the rope haulage systems and the pulleys and the problems with water and pumping that I'm particularly interested in. And there's a pit near me at Astley Green just off the East Lancs Road where they've got an engine that is definitely worth going to see. The pit itself has been closed for years now, but a lot of it has been left standing and there's a group of enthusiasts who are trying to restore the pithead machinery. What a task they face, though. I know how much time it takes for me to get my steam engines done, but a whole pit, well, that's something else.'

Fred had visited Astley Green on many occasions over the years. The first time had been in the 1970s when it was still in operation. 'I went into the engine room,' he said, 'and I must say it were quite a fantastic sight watching the engine revolve at great speed, raising and lowering the cages in the thousand-yard-deep shaft. Astley Green Colliery were a good 'un. I've got lots of mining books and some lovely pictures of the winding engine at Astley Green being put up. It's so big they must have built the engine house around it. And I've got a picture of the men who did the job, all standing there big-capped and very proud. When they'd done it, those men must have had an unbelievable sense of satisfaction as they watched the engine go round.'

Fred was fascinated by all the machinery and by the hustle and bustle of the colliery on the surface, but he couldn't wait to work his way in with somebody who could get him under the ground. Once he'd been underground and looked around he got even more interested in mining and particularly in the old ways of mining. Some of his happiest times were spent sitting in the pubs and clubs near old coal mines talking to ex-miners and the older they were the better because they could tell him what it had been like when they first started.

Just after the Second World War coal-mining employed more men than any other industry. When the Labour government had nationalized the industry in 1947, it still employed 750,000 people. But the decline of industries like cotton helped to reduce demand. At the same time the discovery of North Sea gas and increased imports saw consumption of domestic coal fall even further. Pits began to close down at an alarming rate, leading the miners to a series of strikes. It culminated in the year-long dispute which finally ended in 1985. Some observers argue that the strike effectively signalled the end of mining as it had existed for so long. For others, the final blow came when the Conservatives shut thirty pits in 1992. Trade Secretary Michael Heseltine called it the hardest decision he had ever had to make. Of all the traditional heavy industries, it is probably mining which has suffered most. It now employs just a tiny fraction of the huge numbers who once toiled underground to fuel the country.

Like the Lancashire cotton mills, the pitheads and the winding gear have either disappeared or are now standing deathly silent. By the time Fred went for a second visit to

Astley Green the mine was closed and semi-derelict. 'The scrap men had been,' he remembered, 'and all the windows in the engine house were broken. The rain were going in at one side and out t'other and they'd nicked all the brass off the engine. It looked very sad and only beat the scrap man by the skin of its teeth. But now, thanks to the work of the enthusiasts who came along to rescue it, it's almost ready to run. They've got a Lancashire boiler for it and they have a grand collection of odds and sods. They've also got a steam hammer, which I've had a bit of a play with, and about five or six steam engines, all connected up to a big vertical boiler.

'The trouble is that's all that is left of our pits now – tourist sites where you can see a bit of our lost industrial heritage. Even though there's still coal in them we've lost nearly all of our pits. I think we've lost them because of the ease of drilling holes in the North Sea and getting that invisible stuff up. The government must know things we don't know about the abundance of gas and that is why they shut all the coal pits. If there were any thought of it running out within our lifetime they wouldn't have shut the pits. I think it's very interesting; it was coal that made the Industrial Revolution and it was the Industrial Revolution that made Britain great. We've shut all our coal mines down and what do we do now? There's a lot of brains around now and we are world leaders in computers, so they tell me. But all that's left of coal-mining is part of the tourist industry. I think that is sad.'

In the programmes we made with Fred we visited all of

Britain's major preserved collieries and Fred did a great deal to raise awareness of coal-mining and the part that it played in our history. Jim Daykin is mine manager at the National Coal Mining Museum and when we went to film there recently he told us what he thought Fred's contribution had been. 'Fred had a real enthusiasm for mining. Up until his series of programmes, people probably saw mining areas and mines as dirty places where people would turn up for work and work twelve hours down the pit and then go home and sit in a tin bath in front of the fire. I think what Fred did was to raise people's awareness of the fact that these mines were actually fascinating and that there is an enormous amount of industrial heritage and archaeology to be seen in the remains of these places. The other thing that Fred brought over was the comradeship amongst the people. It was people who made this industry what it was and Fred's programmes didn't just focus on the archaeology, but brought to life some of the people and the jobs they did.'

John Jacques, an ex-miner, is now a guide at the museum. He'd met Fred on a number of occasions when we filmed at the museum and was always impressed by his understanding of mining engineering. 'As a miner,' he said, 'I respected Fred's knowledge and the fact that he could grasp the principles of situations and machinery that he wasn't particularly familiar with. I think he raised the level of public awareness about our mining heritage tremendously and of the contribution of the miners to industrial history, especially for people from non-mining areas. That was very important.'

Fred made friends with many ex-miners and mining experts. Alan Davies was curator of the Lancashire Mining Museum from 1985 until it closed in 2000 and Fred turned to him for some advice before starting work on his replica coal mine. Alan recalls how the idea gradually emerged that Fred would like to have a mine of his own. 'Fred would have a go at anything,' he said, 'so I thought I'd see what information I could get for him on the various structures that you'd find on the surface of a colliery. One of these was obviously the pit winding with the head gear and Fred said he needed an illustration of a traditional wooden pithead, about thirty feet high, that he would possibly be able to duplicate. I tracked one down from a colliery at Blackrod near Bolton and when I showed it to Fred he was very excited about it. "Oh yeah," he said, "I can make one of those." And off he went and made a start. So initially it was just the head gear and then the idea came later of why not sink a shaft beneath it.'

When he built the head gear, Fred went about it in his usual way. 'To actually build a pithead frame,' he said, 'is no mean feat. There's a lot of big iron bolts in it, eighteen inches long. If you had to buy them you would pay a lot of money for them, but just through scrounging and circumstances I ended up with enough materials. One of the hardest things to get was the wheel at the top because, of course, they no longer make pit wheels. To find one that were of suitable proportions for the job were fairly difficult. After a bit of searching round, though, I managed to get all the ingredients together and I got a surge of enthusiasm and made the head frame. I laid it horizontally on the floor at first because I were really, really scared of putting it up, because the council might come and tell me to take it down. Anyway I decided, or at least a friend of mine who had more brains than me decided, that we should do a simulated photograph of this head gear standing up in the garden and take it round the neighbourhood and show it to the neighbours and say would you mind if I erected this head gear in my garden without mentioning anything about digging, because if you mention digging everybody starts panicking.

'Anyway we then applied for planning permission and a nice gentleman came from the planning department. He didn't say a lot at the time, but he went back to the town hall and apparently what he said there were quite nice and complimentary. He said, "I have examined the partially constructed pithead frame and find it has been tastefully crafted from timber and tie rods and iron plates and bars and I feel it would be an interesting asset to an already interesting back garden." But then the planning committee had to come down and have a look before they could pass it. There were twenty-five ladies and gentle-men from the planning department who came down in a motor coach and looked at it while it were lying on floor. Anyway, when they'd seen it and they read my petition they gave me permission to fix the thing up.'

Once he'd got the head gear in place Fred started to do a bit of research into the feasibility of the serious business he had in mind of digging a real mineshaft under it. He got hold of the geological survey maps from the 1840s and found that the house and the garden were on the site of a tip. At the time there was a big ironworks in the middle of Bolton and all the cinders were tipped down this side of the valley.

But Fred wasn't able to start the digging for another seven years. There were always too many other things to do. Then, at a meeting at the BBC in 2003 when we were filming *Age of Steam*, I mentioned Fred's pithead gear and his plans for a working coal mine in his back garden. It was soon decided that it would make a good subject for a one-hour documentary in which we would follow Fred and his mates as they dug the mineshaft and built the surface workings. At the same time we would go with them to some working collieries and some of the heritage mines to look at things like tunnelling and pit props and all the engineering that went with a mine. The film would be a bit of a return to the old style of programmes that Don Haworth had made with Fred. We would follow Fred around, shooting things as they happened in the garden. Fred wouldn't be the presenter so he wouldn't have to remember any lines. He was going back to being the subject of the film.

When I told Fred he was delighted. With a film to be made about his mine, it gave him the perfect reason to divert a lot of his attention to the project that had lain dormant for so long. As with so many things in Fred's life, television had set the agenda. But being able to spend so much time in his garden while we filmed was perfect for Fred. Sheila Dibnah explained the significance of the garden and the reservations she had initially about the mine project. 'Fred's garden was his world, he was very passionate about it and I think it's fair to say that the only time Fred was truly happy was when he was in his garden. I think he'd got so far along with everything that he felt satisfied that he'd accomplished what he wanted to do there, like making his steam workshop. But there was always something else, like in the last year of his life, when he started wanting to dig his pit. And it's a great pity he didn't get the chance to finish that.

'I wasn't very happy about it, I must admit, when he came up to me with the idea of digging a hundred-foot hole in our back garden. In fact I was absolutely terrified because it's only seven feet away from the back door and I thought I might find myself in a position where I woke up one morning and the house was sliding down the hill. But of course, Fred had all this great engineering capability and skill and he had many friends who were ex-miners. I could see how important it was to Fred to do this and I thought, who am I to complain. So after a while I did go along with it, and as time went along, I began to think that it wasn't a bad thing. And in the end, I think that's what kept him going so long, that and working on the engine.'

When we started filming in the late summer of 2003,

Fred had made a start on the shaft. He'd dug a hole about two feet deep, all lined with beautiful brickwork. 'We started in secret really to sink the shaft. It's lined with nine inches of brickwork which is just ordinary practice in mineshaft sinking unless you meet bad ground that's full of water and you have to resort to mechanical means to keep it back like cast-iron lining, which is what they did in the olden days. Modern shaft sinking is all done with concrete, but we're doing it the traditional way. It's all headers nine inches thick, which leads to a very neat finish on the inside; nice and round. When it were first done all new brickwork in a mineshaft looked really good, but now if you look at a proper old mineshaft it has watercress growing up the side. But the next important thing with this one will be to make a couple of doors with hinges that close over the hole. This is in case we kick a spanner down on to the lads below.'

As we started filming Fred explained the grand vision he had for a whole replica colliery. 'At the bottom of my mineshaft there will be a tunnel to the river down in the valley and there will be an endless rope hauling system running down the hill to connect the entrance to the tunnel down by the river with the top of the mineshaft. We could have a really interesting one because we've got the embankment which rolls down at about forty-five degrees for a hundred feet. The best way I can explain it all is along comes the cage when it's wound to the top of the shaft with two tubs of coal in; they crash into ten tubs of coal on the landing plates with bells continually ringing and the wheel going round. As you see the tubs of coal coming up you would have the immediate thought that they're very busy down there today, a lot of men working hard down there digging coal, but really it would be just the same two tubs of coal going round and round and round up the mineshaft, down the embankment on the rope haulage railway and then along the tunnel to the bottom of the mineshaft. It would be, without a shadow of a doubt, a very interesting piece of engineering that all the other preserved pits and mining museums haven't got, which is a bit sad.'

The plan certainly involved some engineering on a grand scale, but Fred was confident he could do it, especially with the helpers he got together to work with him. 'We've got all the ingredients,' he said. 'Two good men who spent all their lives down the pit from being fifteen years old until they shut the pits and they've even got their tickets for being under-managers. They both retired long ago but they are still great mining enthusiasts and are as keen as I am to get the mineshaft completed. We've got Alf, who spent his whole life down the pit wiring tunnels and sinking shafts and messing about like that right up until the demise of the Lancashire bloody coal industry, and Jimmy is exactly the same, a lifetime of digging and engineering and maintenance work down the pit. Then we've got Ian, who is our chief stoker, and another lad, Bill, who's been very successful. He has a Rolls-Royce and owns Westhoughton Labour Club. He were a contractor down the pit; that's a man who does all the nasty bits, the risky bits and all that. He must have had some dodgy times, so he knows a thing or two about safety underground. Then there's a man who teaches the fire brigade about safety and gas and that sort of thing and another lad who's a surveyor working for United Utilities. We're not going to do anything very dangerous and my fine team of men know how to look after themselves, so really I can't see us coming to any harm.'

When filming began, digging the shaft started in earnest and they soon got down to about twenty feet. At this stage Sheila still had reservations about this big hole that was being dug just a few yards from her back door, but she recognized Fred's passion for it and for coal-mining in general. 'I'm not quite sure where his interest in coal mines came from, but I think the basis of it must have been because of the sheer skill and toughness of the miners. It was almost a glamorous profession to Fred and to him miners were heroes. He had a great respect for anybody, such as Alf Molyneux, who had been a miner in the past. To Fred that was a definition of a proper man. Somebody who could do something that required great skill, tenacity and being a hard man. Those things were important to Fred. He had no time at all for anybody who would just sit behind a desk and push a pen, as he would say. He couldn't understand the concept of a man wanting to come home from work and cook his wife's meal. I think he must have made me about two cups of tea all the time we were married. I didn't mind though because he'd have probably ended up making meat pies, and with me being a vegetarian it wouldn't have gone down very well.'

Alan Davies would come down on a Saturday morning when he wasn't working and he described the *modus operandi*, as Fred used to call it. 'He had a bit of help doing the mineshaft, because some of us who were friends had worked in mining so we had a sense of the potential dangers of methane gas emerging from the rocks and the fact that you didn't just climb into a hole in the ground without doing any gas checks. So Fred knew he had a sensible bunch of people with him. They were also fairly strong people who would do all the hard work, while Fred oversaw operations. Occasionally he would get in and swing the pick a few times and then climb out. But usually, because visitors often arrived unannounced, Fred had to be on duty when somebody walked down the drive for a chat. I think Fred liked to see himself as the colliery manager with his waistcoat on, overseeing operations, and we were his miners in a sense.'

Fred was very much the boss in his own garden. In the last couple of years of his life, Alf Molyneux used to be round every day working with him on the tractor and on the mine. He recalls one day when Fred was working at the top of the head gear. 'I remember he was up the ladder and he said, "Alf, can you get a hammer?" So I got him a hammer and took it up the ladder. I'm not so struck on ladders myself but I got up to him and gave him the hammer and he said, "That's not the one I wanted," so I said to him, "Why the bloody hell didn't you take the hammer up with you?" He said, "Listen, that's what you're here for. Every good tradesman has a good labourer." '

As they dug they came up with evidence that backed the research Fred had done on the site. They started to dig up a lot of old bottles which had been dumped there by a mineral water company when it had been a tip. 'Where wc are digging now is definitely not virgin land,' Fred said. 'What we're digging through at the moment is the Victorian tip. I think we're going to have to go maybe another twenty-odd foot before we hit the natural valley which, of course, is sand and pebbles and a bit further down we will come on to coal measures. But we're not mining coal. We're just digging a bloody hole. Further down there's going to be rock outcropping into the river. As we dig we don't really know what we are going to hit. But when we do hit rock we have drills and a steam-driven comprcssor. So we've got all the ingredients; we've got the plans, we've got everything we need and it's all perfectly safe. I mean even my house is quite close to it. My wife initially kicked up a bit of stink because she thought, being a woman, that I just wanted to dig a hole without supporting the sides and she thought the garden would disappear and the house would disappear, which is a load of nonsense. As far as I can see, every-thing we've done so far conforms to mineshaft-sinking safety regulations and I would welcome the coal mines man or the inspector of mines to come and have a look at it and to have a chat about any other safety aspects. It's all taking shape slowly but surely and anybody passing here who didn't really know would think blooming heck, they would think, it's a proper pit because it looks right. People actually come in and say, "Where did you find that pithead gear?" and I tell them, "I used to be a joiner. I made it myself." '

When Fred and his team were not digging, there was work to be done on the steam winder. And during this hectic time there was still plenty to be done on the tractor, so the garden became a constant scene of activity. By now Sheila had become used to it being an industrial site. 'A good day's work was Fred's work ethic. He said that to me many times. I consider I've had a fairly busy and interesting life, being in showbusiness and the entertainment industry when I was younger, and I have travelled the world, but to Fred that all meant nothing. For Fred the fact that I haven't got my hands dirty and been down a pit somewhere meant I hadn't done anything of any real worth. For Fred it was all about hard graft. If I'd have showed willing and gone down that mineshaft and shifted a load of earth, then I'd have been in his good books. But of course I'd go out there and say, "What have you been doing today, Fred?" and, "Oh, is that all you've done?" and Fred didn't like that. He could be difficult at times, particularly in these last years. I think you had to be a strong person to live with Fred. He wasn't somebody who would suffer fools gladly. If he thought you weren't interested in things that he was interested in, he could be quite rude. I always remember one of his favourite video recordings was about the electrification of Bickerstaffe

Colliery. I must have sat through that archive footage maybe ten times before I got fed up of it!'

Fred was now working like a man possessed. Not only did he want to get his mine finished, but there was the tractor as well. And he knew that time was short. His whole life was ruled by these twin obsessions and inevitably it was taking its toll on his relationship with Sheila. As with his previous wives, his passion and enthusiasm and the sheer bloody-mindedness that drove him to get the things done that were important to him meant that he wasn't always easy to live with. When this was allied to the sense of bitterness he was feeling about his life being cut short it created a sometimes volatile atmosphere which inevitably led to rows. And Fred could never hold back from telling anybody who cared to listen all about them.

Most of the friends who were around him at this time would lend an ear to his stories but, as older men who'd been married themselves for many years, they knew that there were two sides to any story and that it takes two to make a marriage work. But whenever there's a group of men together and one starts complaining about 'her indoors', as Fred was doing, there will always be one or two who'll say, 'You want to get rid of her.' In Fred's case it was no different and that sort of comment just served to pour oil on already troubled waters. But those who were closest to him knew that this wasn't what he wanted. He'd been through the misery of being on his own twice before and he certainly didn't want to spend his last days like that. So he'd talk about the bust-ups he'd had but then he would invariably say, 'But she's not so bad really. We can sit down and have a glass of red wine together and I'll say to her, "Come on, cock, we shouldn't be falling out like this." '

Work on the mine went on. Each time we went over to do some filming there was a different project on the go. The next one was to make the doors to cover the top of the shaft and build a nice engine house. 'We've just latched on to some Georgian sash window frames for that,' said Fred. 'This winter we'll maybe clean it up a bit and put some paint on it and some nice ornamental lubricators on it, so it'll look the part. We're also collecting tub wheels so we can make our own tubs. To do this we've got our own saw mill, just like they had at most pits; we've got two boilers to make the steam to drive the engines; we've got the fifty-foot-high chimney and the thirty-odd-foot-high wooden head gear; we've built the pit bank and we're about to start putting the roof on the winding house.'

Jimmy Crooks was the other main member of the team. 'I think he just wanted to reproduce something that nobody else had,' he said. 'We started off with the head gear, then we started digging the hole, and then we got the steam winder going. Before we got it going properly we used to have to winch everything up the shaft by hand. His grand plan was to have tubs going up the shaft and running down the hill into the tunnel at the bottom and then back up again. I don't think we would ever have got to that stage, but we were doing all right till the council came and put a stop on things.'

Fred being Fred, it was inevitable that the council would find out sooner or later. Up until now he had been getting on with the work in secret. But Fred wasn't very good at keeping secrets. He always had to tell everybody he met about his latest projects and the mine was no exception. The trouble was he told so many people that it wasn't long before word got to the council. Fred realized that he'd been a bit indiscreet. 'I think I opened me mouth to the wrong man,' he said. 'He had ideas, because he was some sort of a mining surveyor and knew people on the council. When he told them they sent me the application form for planning permission and I could see there would be a few raised eyebrows when I submitted the proposal. It is a bit of an unusual one, after all. I thought the council might come up with the "Oh, it's too close to the highway," or something like that. But I could go round Bolton and show examples of mineshafts that are much closer to the highway.

'After the council found out, two gents came along and said, "How do you do the brickwork upside down?" I explained how it was all done but they advised me not to dig any further until I'd submitted the planning application and it had been approved. I asked him why, what's the rules on digging a hole in your garden, and he actually said to me he didn't know. The planning department at the town hall certainly didn't know what to make of it. They didn't seem to be able to find any legislation to say you can't dig a mineshaft in your back garden. Maybe it was something for the Mines Inspectorate.' Fred couldn't understand why he needed planning permission or why there might be a problem about his mineshaft. It was his garden and, after all, the job was being done properly, just as it would have been done if the shaft was being sunk for a real mine in years gone by. And if it did cause any subsidence, it was only Fred's own house that would be affected.

But before he could continue with any digging Fred knew now that he would have to have planning permission. He decided on his tactics. 'What I intend to do,' he said, 'is to do a nice drawing of a section through the side of the valley with the road, the house, the pithead gear over the shaft and all the other stuff that's around there. Once I've got as much down on paper as I can, I'll send it to them and that's when the fat will hit the fan as you might say. There will be a lot of worried people. But I can't see what there is to worry about; it's a simple operation. I mean if it were eight hundred yards deep you could see why they might worry a bit, but this thing we know now is only ninety feet, which is nothing.

'I must emphasize as well that this is a simulated mine – it would be naughty to dig coal. It isn't a real mine, we're not removing any mineral from under the ground. We're just sinking a shaft. It could be a well or a bomb shelter. This is just for fun; it's just a model on a full-size scale. It's not for the public use either. This is for my friends and my entertainment as you might say, because we're all coal fiends and colliery buffs. We all go down disused coal mines and everybody's still alive after being in terrible places where you shouldn't really be.

'I can't see there's any need for planning permission to build something like the inclined plane railway that nobody can see. It's on my land and you cannot see it from the road. You cannot see down the hole either and it will have a beautiful Victorian reeded-column handrail all round the top so it will all look very nice. Then from a safety point of view, when it's not in use the cage will be hanging mid-shaft so nobody will be able to come in and mess with it, and if the fire's not lit, you can't work the machinery. It's not as if anybody could creep in and get it going by turning the electric on; you've got to light a bloody fire and then it takes a day and a half to get steam up. Then there's the doors on top of the shaft to stop any illegal entrants falling down the hole; not like when I were little and there were disused collieries all over the place, and lots of little lads fell down mineshafts when they were birds-nesting, which were all a bit sad.

'As far as civil engineering's concerned I can't see owt that's impossible. There might be some clever power somewhere who says, "You can't do that because it's dangerous." But everything's dangerous. There's people jump out of aeroplanes with parachutes and that's bloody dangerous in my opinion. Going down potholes is another unbelievable dangerous thing, and playing golf. I mean you could get hit on the back of the head with a golf ball or you could be sat having your tea and a golf ball comes sailing through your window. But there's little chance of us doing any harm to anybody else. I personally can't see any problems anyway if you do the brickwork good. But you've always got some guy somewhere who suggests you can't do that because it's dangerous. I mean now you can't even smack your own kids, can you, because they say it's dangerous. It's time they had a bit of discipline back. In fact I were doing a talk t'other night and I mentioned National Service and everybody cheered, you know. They should never have stopped it; it would do them all good.'

But the council didn't see things Fred's way and his planning application was turned down. Jimmy Crooks, who'd worked with Fred on the mine from the start, was disappointed but not really surprised. 'I don't think we could have ever ended up doing it like he wanted it,' he said. 'I don't think it would have worked, not the way he had it in his head. But you couldn't change him. You'd try and explain to him that you can't do it that way or that something wouldn't work but he wouldn't take any notice. So we just let him get on with it. We had some good fun, though, and some good laughs on the mine and on the tractor, especially at dinnertime when you'd get Fred telling his tales. We always had pies, but if Sheila ever came out when he had a pie in his hand, he'd always throw it away or pass it to one of us. I think she was trying to teach him to be a vegetarian.'

Fred wasn't going to give up his vision too easily, so, as soon as his planning application was turned down, he put in an appeal. And Fred and his Dad's Army were not going to be idle. There was still work to do on the traction engine before it was ready for the road and if he could finish it in time there was the possibility of another series in which Fred would travel round the country on it. It was something that Fred had wanted to do for a number of years, so the disappointment about the mine was tempered by the prospect of this great journey.

14

The Last Chimney

The traction engine that Fred had been working on for so long was a time capsule of Britain's industrial past. Every part of it had been invented and made in the United Kingdom. It had taken Fred years to restore the engine and now he was nearly ready to take to the road with it, travelling the length and breadth of Britain in search of the skills and technology that had made it possible to build an engine like this in the first place. The journey was going to be made for a new twelve-part television series *Fred Dibnah's Made in Britain*.

The idea for the series went back to 1999, when I'd asked Fred if he'd be up for the idea of a journey round the country on the traction engine if he could get it finished in time. Failing that, he could travel round on the roller. The series I had in mind initially would have seen Fred and the engine visiting interesting places and meeting interesting people. It would include castles and cathedrals and country houses as well as steam and industrial heritage sites. We had just finished filming *Magnificent Monuments* at the time and I was looking for a follow-up. Fred was keen on the idea from the outset. He'd enjoyed all the filming we'd done for the first two series and all the places we'd visited and the idea of combining this with a grand tour on one of his engines seemed perfect to Fred.

Throughout the following year the idea was developed further. The BBC wanted something a lot more focused, so we came up with the idea for a journey on which Fred would go in search of all the things needed to build an engine like his in the first place. It would uncover and celebrate the remarkable achievements of the craftsmen, engineers, inventors and industrial workers without whose endeavours his engine would never have been possible. Fred's passion was for the age when the words 'Made in Britain' were stamped on steam locomotives, traction engines, mill engines, pumps and pit winding engines all over the world. The expertise that was required for Britain to achieve this pre-eminent position was wide-ranging, from mining and smelting the ore needed for the raw materials to the specialist craft skills of making parts like pressure gauges, gears, valves and bearings. Many of the skills, like boiler-making, had still been around in the 1970s, but a lot of them were disappearing and the idea was that Fred would travel around on his engine to find some of the little workshops where they still survived. For Fred it was perfect. It would give him the impetus he needed to devote his time to getting the engine finished and for the series he'd be able to spend a few months on the engine, travelling round to meet up with a lot of his mates from the steam world.

But it wasn't to be, or at least not then. It was the time of Simon Schama's *History of Britain* and what the BBC wanted was a series on British building, so *Building of Britain* was commissioned and the idea for Fred's grand tour on the engine had to be shelved. It was not until the end of 2003 that it came up again, just after we had made *Dig with Dibnah*, and it was the emergence of Alf Molyneux in that film as the perfect foil for Fred that helped to get the series off the ground. Fred would need a steersman for the journey and Alf had been very good on camera with Fred in the mining programme. With Alf on board as Fred's mate for the journey, the series was given the go-ahead.

Fred was delighted. Since I'd first raised the idea with him four years earlier, *Made in Britain* had become the series he most wanted to do. Finishing the engine after so many years and setting off with it on a grand tour of Britain would be a highlight of his career. There were not many more steeples left to climb or chimneys to demolish, but now, with all the expenses of getting the engine finished and travelling around the country on it covered, he would be able to achieve what had become one of his last great ambitions. The other one, getting his mine completed, had been temporarily thwarted, but the prospect of this new series made up for his disappointment, as did the other good news he received at this time: he was awarded an MBE in the New Year Honours List for services to heritage and broadcasting.

It was early 2004 and Fred knew he had only a short time to live. The series he'd wanted to do for years had been given the go-ahead, but would there be enough time? The engine still wasn't finished. It would need another few months' work before it was ready for the road. The journey itself and the filming, we estimated, would take at least four months. Fred was determined to do it, but he would be getting weaker all the time. His friends rallied round and the garden became a constant scene of activity as they worked to get the engine completed as soon as possible.

Fred was called back to the hospital and told about two new drugs that were being tested. Would he, they asked, take part in the clinical trials for them? It sounded as though it might be a lifeline. Every time I saw him Fred would tell me about some miracle new cancer cure he'd read about in the paper. Could one of these be it? The tests he'd been invited to take part in would take up most of the summer and would involve frequent visits to the hospital. Fred wasn't too keen, especially as the doctors had emphasized that the tests were unlikely to be of any benefit to him. He would be taking part in the interests of medical research that might benefit cancer sufferers in the future. Fred decided against taking part in the trials. He was determined to steam ahead with his farewell to millions of fans even though he was critically ill. It was, he said, something that he was desperately keen to do. He knew it would have been a difficult project, even for someone in the best of health, but he was absolutely insistent that we should go ahead. It was what he had wanted to do ever since I had first raised the idea with him five years earlier.

I was uncomfortable with this. I sat at Fred's kitchen table and tried to reason with him. If there was the slightest chance that the drugs could give him even six more months to live he should take part in the trials. I felt we ought to postpone the filming until he'd at least given it a go, but Fred wanted to get on with things as quickly as possible. Any delay now, he felt, would make it too late. I said I'd ask one of our researchers to find out a bit more about the drugs. The results were not encouraging. They backed up completely what Fred had been told. Both the drugs were in the very first stage of human trials and it was unlikely that there would be any benefits for those taking part in the trials. When I reported our findings Fred seemed almost relieved. We were going to go ahead. Everybody close to Fred at the time assured me that this was what he really wanted. Finishing the engine and making this journey would give him a focus. It would keep him going and keep his mind off the end. It was now a question of getting it done so that Fred could achieve one of his last wishes.

We started filming in the garden as he worked on the tractor with Alf and Jimmy. Fred told us more about the saga of building it. 'When I bought it twenty-seven years ago I didn't really think it would take as many years as it has done to get it to this stage. The original were absolutely rotten. So I started off by making a new boiler barrel. I didn't actually roll the plates but I did all the riveting. When I look at it now and see what I've done, I've almost made a new one. There's practically nothing left of the original.

'When I first got it I said to Don Haworth, "If you give me eight grand, I'll be able to get it finished in six months." Buying the engine appeared in those early films; doing the deal, having a beer, dragging it in the shed and the irate wife, never going on holiday. Then Don Haworth filmed a bit of the beginning of the riveting, but not a lot. From that time on the whole job has been one setback after another. Even the firebox was welded on and the people who did it welded it in places where it didn't need welding so it caused me a lot of bother. I had to grind a lot of surplus metal off that I needn't have done if they'd done the job right in the first place.

'You've got to have a laugh though; it has meant twenty-seven years and two divorces for me. Any man who has got a steam engine will tell you that you've got to have an understanding woman; otherwise you're in big trouble. It's really hard to reckon up the actual time I've spent on it because there can be a period of six or seven weeks when you're working away from home and there's no time or bother with it. You get people writing these articles on how I did my traction engine up and they say things like "I spent twenty-nine thousand hours on it," but it's a bit weird how they can get it dead right down to the exact hour.'

Whenever Fred was asked why it had taken him so long he always had the answer. 'As soon as you light fires in boilers,' he always used to say, 'they seem to have some fatal attraction for people. They just come from nowhere. Then, of course, you can't stop talking, can you? You've got to talk to 'em, so I would say about half the twenty-seven years has been taken up with talking and not doing the job!'

But there was also the question of cost. 'Sometimes,' said Fred 'you go and get a price for something and it's an arm and a leg and you think it's going to bankrupt me. So you chew it over in your mind for three months and finally you come up with a solution. Necessity is the mother of invention and I've saved myself literally thousands of pounds. If I'd had to pay for a lot of things it would have been a fortune. For the rubber tyres alone you are talking a thousand pound on each wheel if you pay for it at Dunlop's. But there's ways and means that you can find these things and get them practically given to you. Various benevolent people have been very kind to me.

'The actual original boiler barrel has grass growing out of it now. I had to make a new one from scratch. Everything was going very well; we had almost got to a Meccano set stage when all you have to do is put all the individual bits together, really easy, but disaster struck. Unfortunately, when we'd just got six rivets left to do, a horrible crack appeared in between the four rivet holes where the barrel goes into the throat plate. That was really bad news, because this is the most complicated piece of plate on the whole boiler. The initial thought was that it was just a pencil line, but it wasn't; it was a proper crack in the iron. It was most likely to have been caused by the way the thing was made in 1912. The only solution was to get it welded. So that's what I did. I called the welder and while I was waiting for him to come, there was a terrible disaster in America. A traction engine blew up, killing five people and injuring forty. I thought if that was to happen in England it would be the end of the traction engine movement, they'd not allow us out on the roads any more.

'While I was waiting for the welding to be done, I thought of having the engine radiographered, or X-rayed. A man came with his television sets and his probe and his gel and he went round the seam of the rivets. As he did it his face was getting longer and longer. All in all there were twenty-seven cracks in the plate, he said, invisible to the naked eye. This meant we would have to make a new throat plate, which would be quite a feat. I succeeded though.'

By this time Fred had a lot of friends helping him, and once the new throat plate was made progress was rapid. Alf Molyneux and Jimmy Crooks were among those who got involved. Alf says, 'I met Fred and we got talking about pits and he said, "Oh, I've got a pithead gear in the back garden," so I said, "Oh, I'd like to see that, being an ex-miner." So I came down to have a look. But I came on the day when he needed a bit of muscle power on the traction engine and he said, "Can you just give us a lift with this?" and that were it, I were trapped.'

One of the biggest problems Fred had faced was finding places that could supply him with the parts he needed. This was particularly true of some of the smallest parts, like the nuts, bolts and rivets to hold the engine together. The little workshops where things like this were made, which almost every town once had, have now all but disappeared. 'In this modern and advanced age that we live in,' Fred said, 'it gets more and more difficult to find people who can actually do work like this. It's incredible really; there's a heck of a lot of bodge-ups and you can't really afford that sort of thing with a potential bomb, and that's what an engine like this is really. If I had to start again now to do what I've done on this loco-motive boiler I'd be in trouble. The people who've helped me and done various bits of machinery for me have nearly all gone.

'The gearing, all the cog wheels and all the trans-mission is brand new. When I first looked at the gears, they didn't look too bad, but gears are funny things. If they're a little bit worn they make a lot of noise and I don't like too much noise. So I ended up with a brand new set of gears. Then there was the tender; it was a bit like a patchwork quilt. I've never been one of those men who like things patched up so I decided to make a brand new one. With the various methods of getting plates red hot and bending them round bits of iron I managed to make a brand new tender. All the bearings are brand new, the cylinder block is new and it's got new piston rods, new valve rods and all new pins. When it's screwed together I think it'll be actually better than what it was in 1912.'

Before all the bits were screwed on, however, Fred was able to fill the boiler. He got it up to 175 lb per square inch and there were no leaks, which was pretty good, he said, 'considering it is a home-made boiler, made in my back yard. It looks very rusty at the moment, but that's good for it. Leaving it standing like this helps to seal the rivets up. Very shortly, though, when we've got it up to the hydraulic test pressure of 300 lb per square inch, we'll be able to shot-blast it, paint it matt black, then roll it into the shed and start to assemble it. I'm really, really looking forward to that.'

In the later stages, when it was being assembled, Alf Molyneux used to go to Fred's every day to work on the engine. 'That was great,' he said. 'I hardly knew what a nut and bolt was when I started helping Fred but his enthusiasm rubs off on you and you learn as you go along. He was always a very easy chap to work with, very easy-going and I've learned a lot off him. He was a good teacher. But he used to teach himself as well. I watched him once or twice when he came across something he'd not done before and you could see he was struggling. You could see he were thinking about something so you wouldn't disturb him. Often he'd read a book in the evening about some mechanical problem he'd come across and then he'd have a go at it the next day. The other thing was he was never afraid to ask. He might be in a factory somewhere and he'd talk to mechanics and he'd ask their opinion about a problem he was coming up against, or he'd ring one of his steam buddies and ask them.

'One of the things that always came across when working on the engine was that he was a perfectionist. You'd put a rivet in and if it weren't quite right, it was drill it and get it out and they take some drilling and getting out, I can tell you. But if you did do something wrong he wouldn't start shouting and bawling about it. He'd just say, "No, that'll not do," and you had to do it again. So you quickly learned only to do it once and do it right.'

Just before the engine was completed another job came up. With chimney demolitions now few and far between, when Fred was offered a felling job he wasn't going to turn it down, however busy he was. 'It's two years now since we did the last 'un. They don't come that often these days and every one that does come up now is a really big occasion.' Fred had demolished eighty-nine chimneys over the past forty years. His last chimney, as it turned out to be, was at the Park Mill at Royton, near Oldham. The Park Mill stopped spinning cotton in 1962. In 2003 a deal was reached to redevelop the site for housing. The mill buildings were all demolished, leaving just the three-hundred-foot chimney still standing. The felling was scheduled for lunchtime on Sunday 9 May and a big crowd was expected. Fred got the old team together and got all the old tackle out of the shed ready to start preparations on the Monday before the scheduled drop. When they arrived on site, there was a problem. Fred needed a compressor so that his men could attack the base of the chimney with a pneumatic drill. They planned to hire one locally, but they'd forgotten that it was the May Day Bank Holiday, so everywhere was closed. It put things back for a day, but they still had plenty of time to get everything ready for the Sunday morning.

As with all his fellings, Fred would use traditional techniques derived from mining technology. This involved knocking a hole in the base of the stack and then propping it up with telegraph poles and wooden wedges. These pit props were burned in a large fire before the chimney collapsed safely in the right direction. Fred employed a loyal team of helpers for the job. Some of them had worked with him for more than thirty years and for this one, as on so many occasions before, they'd all taken a week off their normal jobs to give him a hand. A chimney like this took a year to build, but Fred had under a week to bring it down. His only concession to the modern world was the compressor, which had now turned up.

When the hole was four feet wide it was time to start putting the props in. It was important they got this right. These props take the weight of half the chimney. That's hundreds of tons. The chimney would fall when half its supporting brickwork had been removed. Fred had got to be certain that it would go in the right direction. To do this, he had to be sure that the hole he was making in the brickwork was in exactly the right place, because that was what determined the direction of the fall.

It was 6 a.m., Sunday morning. Felling day. The end of an era for Royton's spinning industry. Overnight, firewood had been delivered to the base of Park Mill's chimney. The police were there early, blocking the surrounding streets. As the crowds began to gather, Sky News arrived with their outside broadcast van complete with satellite dish. It was a mark of Fred's fame that when he did a chimney-felling now it went out live on Sky News; all good publicity for the developers and for the demolition contractors, who had a huge banner down the side of the chimney with their name on it. Jack and Roger had come over for the event and were helping with final preparations, while most of their dad's time seemed to be taken up doing newspaper and television interviews. In the streets around the site stood burger bars and ice cream vans to cater for the ever-growing crowds and there were more cameras than you would see for a royal visit.

A steady stream of VIP guests was arriving, all wanting to see Fred and to get a close look at the chimney, but as the time for the drop approached Fred began to move them and the reporters and photographers back to a safe distance. It was all very different to the first chimneys that Fred had felled forty years earlier. But once Mrs Dibnah had lit the fire in the customary manner, everything went exactly to plan. Within minutes thick black smoke was billowing from the top of the chimney. The television crews and press photographers were all in position, cameras trained on the chimney. Then right on schedule, just over twenty minutes into the burn, there was a crack and a split appeared in the chimney. Fred sounded his klaxon as he ran away from the base and the Park Mill chimney fell exactly where Fred had planned it. All that was left was a pile of bricks.

It was the end of an era, not just for Royton's textile industry but for Fred. 'All gone,' he said, somewhat ruefully. 'Really, me steeplejacking days are nearly over with.' But being Fred, he was able to put a positive spin on it. 'This of course will enable me to do lots of things that I've always wanted to do. Like now, me tractor's nearly finished and I'll be able to go off on my grand tour of Great Britain, meeting interesting people who do interesting things. Before we set off, though, I want to do a few road tests with it around the local area, and visit one or two people who've helped me with machining and engineering the bits that I couldn't do myself. Should be good fun, I think, especially as Jack is over from the Isle of Man and he's going to come on it with me to steer.'

But getting a traction engine out on the road isn't like getting your car out of the drive. There are, as Fred said, hours of preparation involved. 'You've got to procure the sticks and an oily rag, and get the fire lit, and make sure there's enough water in the boiler. If there's no water in the boiler you're in trouble. And then of course you go round with the oil can. There's literally dozens of oiling points. The whole procedure took a couple of hours before we actually set off. I can't think of anybody who's ever done it any quicker.'

Two hours later Fred was out on the road with the engine for the first time and it was going well. 'Driving a traction engine,' he said, 'is a fairly simple thing really. There's very important things you've got to watch like the water level in the boiler. If you're an absolute raw recruit or a beginner and you put in for your driving test on the traction engine, the man who comes to do the test is happy as long as you can get round a corner, do a three-point turn, and stop the thing. But there's a lot more to it than that. I mean they don't ask you about what would you do if the water's disappeared past the bottom of the glass, which is very important because a steam boiler is basically a bomb. To make it go is what's called the regulator, which opens the steam valve that lets steam into the cylinders or, in the case of the throttle on a car, lets petrol into the carburettor. An internal combustion engine, though, is a very feeble thing compared with the power of steam.'

When he was building the engine, the local scrap metal merchant provided a lot of the materials Fred needed. So that was the first place he visited to let them see the finished engine. On his first day out on the road with the engine he was looking for a metal bar to make a driving pin for one of the wheels. But when you get to the scrap yard you never know what you're going to find, and on this occasion Fred's search was fruitless.

His next stop was the place where he got most of the copper and brass for the engine. When he got there he showed the owner all the bits and pieces that had come from his scrap yard. 'All the metal that these taps are made out of came out of this scrap yard and all the lubricators and these nuts. And there's a tap round the corner that we get the washing water out of, that's come from here. Even these greasers come out of this scrap yard. People throw away stuff that they think is never going to be of any use in the modern world, but people like me can put it to good use in the old world!'

When Fred got back home after his maiden voyage he declared that he was very pleased with the engine. 'Today's been ninety-nine per cent successful,' he said. 'The leaks are a bit worrying but they're not that serious. Then there's one or two odds and sods that need seeing to and then we'll be ready for our world tour! There were some days I never thought I'd ever see it. But I've done it, it's finished; it's running. The only thing about it though is that in top gear it goes very fast, much faster than the steamroller, and I'm afraid to say I'm not quite used to that yet. I nearly crashed through the neighbour's fence with it when I was getting it out.'

Next day it was time for some more road tests and it was Alf's turn to have a go at steering. 'We've got it on the road,' he said, 'and it's beautiful and it's an achievement. I've a bit more love involved in this than the roller because I've helped to put it together.'

On this little trip, Fred was heading for Metric Engineers, where they had machined the gears for the engine. That sort of engineering was just too big for Fred to do in his shed. On the way back home it started to rain. Very soon the engine was steaming through a real downpour and Fred and Alf were getting very wet. Like the roller, the tractor should have had a cab, but that hadn't been built yet and to get it done before setting out on the grand tour would cause too much delay. So the engine driver and his steersman just had to hope for a good summer.

But a bit of a soaking wasn't a problem for Fred. The engine was doing 12 mph. Fred was very pleased with its performance and he was getting used to travelling at these high speeds. 'Basically when you're steaming along the road, if the road's a good 'un and wide enough and everyone's getting by you, you can relax. It's a bit like driving a canal boat. The steering's very similar, a bit this way and a bit that way, but there's no signalling system on it. You've got to hang outside and put your hand out like they did in the olden days.'

The engine was ready for the road. All that was needed now was some last-minute work on the living van they'd be towing round with them. As the signwriter worked on it Fred told us why the van was so important. 'It's like a support vehicle,' he said. 'In it there's the toolbox, the coal, the brewing-up tackle, the beds and everything you really need. When it starts to rain, you've got your own mobile home. You can just stop on the side of road and have a brew and wait till the rain stops.' The traction engine men of old would have been on the road for days at a time and the living van would have been their home. This was the way Fred wanted to do things on his journey. It was certainly the way he would have done it five years earlier when he had still been in good health.

When we were planning the logistics of the journey and how it would be filmed I'd discussed this question of accommodation with Fred. Did he really want to do it all for real and stay in the living van? 'Aye, well, I would,' he said, 'depending on whereabouts we were and the degree of safety for the engine. I wouldn't just leave it at the side of the road. Some engine men in the olden days used to just jump off and leave their engines and go straight into the pub, but I don't want any of that. With the steamroller over the years I've slept at the side of the roadway. You just pull up when you're tired and worn out and call in the pub for a pint or two and then retire to bed in the shed.'

Fred thought the journey would take about four months. 'You don't want to be rushing,' he said. 'There's some quite long journeys involved but it'll be uneventful other than waving to people as they pass by. Then there's always the unplanned; things happen, like getting invited into people's houses, which can be quite dangerous. You end up pouring whisky down your throat. "Just bring it round," people would say, "so we can take a picture of it in front of the house," and then you ended up at a party. I've even been invited into weddings.'

Fred had been reading about the long distances that the traction engine men of old used to cover. 'I got the driver's instruction book of Norman E. Box, Lancashire Steam Road Rolling Company, from Broadheath, Manchester. It actually tells the drivers which town won't let you do what. Sometimes they wouldn't let you go through in the hours of daylight; you had to wait outside the town until it were dark before you could pass through with a traction engine or road roller. Then there was the business of filling them up at horse troughs and which ones could be used. There's other quite unbelievable things in the book, like, in case of an accident on no account must you speak to the police; ring head office straight away. That's incredible.'

Much as Fred would have liked to do it like the traction engine men of old, staying in the living van in a lay-by or outside a pub each night, we agreed that, given his state of health and the demands of a filming schedule, he would be booked into hotels with the rest of the crew and secure parking would be found for the tractor. Then we looked at the rest of the logistics for filming with a traction engine. Arrangements always have to be made with locations well in advance and schedules have to be kept to. But how fast would the tractor go? How much coal would it need and where would we get it from? How many water stops would the engine need? Because it was new and Fred hadn't been out on it before, it was all a bit unpredictable. He'd only really got the roller to go on. 'The steamroller will do about twelve miles on a tankful and then you're at rock bottom – boiler's empty and tank's empty. But this is a different kettle of fish. According to the book, it's capable of doing fifty miles on one fill-up of water. But we shall have to see.'

We based all our schedules on a speed of 12 mph. There were some very long distances between locations which would have taken days and days of travelling time. Even if Fred had been in good health and able to do the journey on the engine himself, it would have been difficult to work a filming schedule round it. In Fred's present state of health it would be impossible. We needed a low-loader to transport the engine around. Fred recommended Alan Atkinson, a long-standing friend and fellow steam enthusiast from Preston, who owned a roller and a steam wagon. His company, Atkinson Trailer Hire, had just the sort of vehicles we needed and Alan had, in the past, transported Fred's roller to various rallies up and down the country. When we asked Alan to provide the transport for the tractor he was pleased to join the team.

The low-loader would cover the long distances between locations, but there would still be a lot of driving for the tractor and getting round under the power of steam wasn't going to be easy. An engine like Fred's wasn't really designed to be driven in modern-day traffic. 'When these engines were made,' he said, 'the traction engine were basically king of the road, and everything got out of the way of it. As time's gone by and we've got into the modern world, you've got to be on your toes because these things don't stop quick; you need a yard or two to stop. A lot of people who own these things don't go on the road; they're scared of actually getting amongst the modern traffic.'

But Fred was looking forward to the trip and so was Alf. 'Oh aye,' said Alf, 'I'm going to be on this holiday with Fred. That'll be a once in a lifetime opportunity. I'm the steersman and I've no doubt he'll show me how to drive the thing properly. Of course I'll have a running commentary about everything off Fred and that's something I'm looking forward to as well.'

15

Made in Britain

It was May 2004. Fred's traction engine had passed its
boiler test and had performed well on its road tests
around Bolton. It was ready for the grand tour. But first
of all there were some local trips to make. Coal would
power the tractor's journey around Britain but there were
no coal merchants left in Bolton so their destination for
the first day was an opencast mine between Wigan and St
Helens. The plan was to get stocked up there and come
back home that night. The following day Fred was going
to visit Astley Green Colliery Museum before heading
for Cumbria the following week at the start of the tour
proper. Each return journey was about thirty miles. Fred
estimated that, even if the engine only did 10 mph, he'd
be able to do each of these trips comfortably in the day
and get all the filming done.

But things got off to a bad start. The coal they were
collecting for the trip had to be stored in the living van,

so that morning the engine was going to be towing it for the first time. But first Fred had to get it out of his garden, and that wasn't easy because there was a problem with the brakes. It was only after a two-hour struggle that they managed to get it out on to the road. By the time they set off, with Alf on board as steersman and Jimmy following in his car as backup, they were nearly three hours behind schedule.

Whichever way you go out of Bolton it's all uphill, so it was an early test for the engine with the extra weight behind it for the first time. And it was a real struggle. The shiny new engine performed like a clapped-out old wreck, labouring up the hill from the town centre at no more than two miles an hour. A long tail of frustrated motorists queued behind. A mile and a half out of town Fred brought it to a halt. Pressure was low and it just didn't have enough power to get the living van up the hill at anything other than a crawl. There was also a problem with the mechanical lubricator, which feeds the oil into the cylinder, so some adjustments were needed.

After some roadside repairs they carried on, but it was a stop-start afternoon with a lot more time spent standing by the roadside getting pressure up than driving. By teatime they were still four or five miles away from the opencast mine and the next time they stopped they met the mine owner. He'd shut up shop for the day and come home for his tea so, even if they reached the mine, they weren't going to get any coal – and there was another hill ahead.

Getting up the hill brought another problem that needed urgent attention. Water, fundamental to generating steam power, was running low so they had to find some quickly. Fortunately Fred spotted a fire hydrant and brought the engine to a halt yet again. 'We've run out of steam,' he said. 'That's a big hill for a traction engine and there's no guts in this at all. I'm not happy with it. I think I'll sell it.' It wasn't a good start.

Fred explained how close they'd been to disaster. 'There's a safety plug at the bottom of the boiler, just above the firebox. If we'd not managed to fill up in time, it would have melted and put the fire out. This would have meant we wouldn't have been able to move the engine from here until the plug had been replaced, and that's a big job.'

As Alf and Jimmy filled the engine, Fred told us about his policy for getting water. 'You are supposed to have some sort of licence for messing about in hydrants,' he said, 'but I think really you're doing the water authority a big favour because these hydrants are very important things. If the water people haven't been round and lifted the lid off for ten years, all the mud off the road washes down the little holes in the top and eventually they end up full of the stuff. Then you arrive on your steamroller and you lift the lid up and it's very unpleasant. You just have to roll your sleeves up and put your hand down into the mud to reach the cockstop at the bottom that works the hydrant. Then when you turn it on you blast all the mud out, so you've actually done the fire brigade and the water company a favour because you've only had a few gallon of water but you've cleaned out the hydrant for them free of charge. To me, any man with a steam engine who is going along the road when it needs a drink should be allowed to get a bit of water to fill up his boiler free of charge.'

It had been a long, hard day and a bit of a disappointment for Fred, so it was time to find somewhere to stop for the night. He found a pub car park that looked suitable and arranged with the landlord to park up there. It had been hot, thirsty work and he was ready for a pint. As he stood by the engine Fred was in a reflective mood. 'There's something amiss somewhere,' he said. 'It should go better than it is doing. But the mechanical lubricator's fallen off altogether now. It's just hanging on top of the boiler. We'll have to get that fixed before we can go any further. And we've hardly any coal left. What we set off with should have been enough to get to the mine easily but we've run out. The steamroller would have gone a lot further than this. We've also forgotten the big firing shovel and the big rake. With all the excitement we've forgotten half the tackle.'

Jimmy Crooks remembers that first day on the road. 'Everything had to be spot on for Fred. I think that's why it was such a big disappointment to him when it wouldn't perform at the beginning. The trouble was we should have had more time beforehand. We should have been working on the tractor instead of digging. But Fred got involved with his mine and he left the tractor to carry on with that.'

Fred and Alf stayed in the living van on the pub car park. It hadn't been planned that way but Fred didn't want to leave the engine in an unsecured car park for the night. While they were in the pub Fred got talking to a local coal merchant, who said he'd bring him a couple of sacks round the next morning so that he'd have enough to get to the mine.

As the next day dawned Fred and Alf rose early to get steam up. The coal man delivered the two sacks and though it wasn't very good quality it would be enough to get them to the mine. By ten o'clock Jimmy and the film crew had arrived and they were on their way. This area had been at the heart of the Lancashire coal field. In 1950 there were seventy collieries around here, employing 55,000 people. But from the mid 1950s the industry went into decline and the pits began to close. By 2004, when Fred was setting out on his grand tour, fewer than five hundred people were working in the mining and quarrying industries combined, and one of the few places left where coal was still being extracted was Crook Hey Open Cast Mine, their destination.

Once they were stocked up with coal the next stop was the Astley Green Mining Museum, one of the few places that survives to tell the tale of coal-mining in the area. It houses Lancashire's only surviving pithead gear and engine house. Alf remembered it well because it was a pit he had worked at for a time. From Crook Hey to Astley Green it should have been an easy journey along the East Lancashire Road, a fairly flat dual carriageway. In all the travel-time projections we made with Fred, the estimate for this sort of journey was about an hour and a half. Instead it took all afternoon with the engine and living van crawling along at no more than 3 mph and having to stop every mile or so. As the afternoon wore on Fred became more and more dispirited. Ever the perfectionist, it hurt him that this machine that he had devoted so much of his life to rebuilding was not performing properly. 'There's something terribly wrong with this,' he said each time he stopped on the busy dual carriageway.

It was early evening before the engine arrived at Astley Green, too late to do any filming. And the long hours spent on the footplate over the previous two days had clearly taken their toll on Fred. He'd not really left Bolton yet to start on the journey proper, but it was already clear that he was too ill to spend much time on the engine. As he climbed down wearily from the footplate he looked up at the rusting pithead gear and talked to Alf and Jimmy, who had spent all their working lives in these Lancashire coal mines. 'It's got to be worth a big grant from the lottery people to keep it stood up,' he said. 'If you wanted to open a ballet school there'd be millions for you, but not to paint something like that, which has kept a lot of families in bread and butter for a long time.'

Parking at the colliery museum was secure so they put the covers on the engine and left it there for the night. They went back over the weekend and got the mechanical lubricator fixed and did a bit more tinkering with the engine to try to get it to perform better.

After coal the other essential material that had to be dug from the earth for an engine like this was iron ore, so Fred was going from Lancashire to the Cumbrian coast to visit the last deep-working iron ore mine in Europe, Florence Mine at Egremont. First thing Monday morning Alan Atkinson arrived with his low-loader to transport the engine. But it was the first time and inevitably it was a lengthy process. It was lunchtime before the low-loader arrived with the engine at the spot in Cumbria where we were due to start filming and well into the afternoon before it was steamed up and ready to go. When he arrived with the engine, Fred had a pleasant surprise. Jack was there to meet him.

'My eldest son Jack's come over from the Isle of Man to give us a lift on this trip. And really, the way we're going on, I think we need all the help we can get. We've got Jimmy and the support vehicle behind us. We're stopping so frequently due to lack of steam and going up hills, and a pair of flashing lights behind is a must on country lanes like this. But while we're up here we're calling on a mate of mine, Mr Richard Ransome, which is where Jack is staying. Dick is a fellow traction engine owner and steam engine enthusiast and we'll be able to do some running repairs in his workshop.'

The main reason for the trouble was that in all the hurry to get the engine on the road, Fred hadn't really had enough time to do all the fine tuning he would have liked, so he kept coming across little problems that needed to be sorted out. The main one was lack of steam and Fred thought he knew the reason for this. 'I think I've put the cylinder block in one of the pistons a bit too far forward. So what I'm going to do when I get to Mr Ransome's is shorten the piston rods by about a quarter of an inch.'

But Fred had to get there first. His engine was a four-horsepower model, built to pull around fifteen tons. But right now it was struggling to pull two tons and the hills of the Lake District were proving a real problem. Long queues of traffic were building up behind Fred, who was worrying about whether the engine would get the living van as far as Dick's. But help was at hand. Dick got one of his mates to come out and hitch the van up to his Land-Rover so it could be towed to his workshop. 'It's a bit of an insult to our engine,' said Fred, 'but at least it's going to get us there. It's been such a disappointing journey. In the morning when it's cooled down we're going to have to take the cylinder end covers off and weigh up how much we can take off the end of a piston rod. And I know Dick'll help me because I've known him a long time.'

That evening Fred, Jack and Alf stayed with Dick and they contacted another old friend of Fred's to come over to see if he could shed any light on their problems. Fred gave me the news the next morning. 'We've got Roger Mallinson coming, who's quite an authority on steam engines. He makes steam engines for rich people's steam yachts, you know, for Lake Windermere.'

Fred had a great respect for Roger's engineering expertise and he cheered up considerably when Roger arrived and gave his verdict. 'Generally it's not in bad condition,' he said. 'I'm a bit worried about the way the piston is covering the port opening in the cylinder so when the valve opens there is no room for the steam to get through. What I propose to do is lend you a high-speed burr so that you can machine a lot of the iron out of the cylinder round the port.'

The engineering work was done and next morning the low-loader was ready to transport Fred and the engine to the Florence Mine at Egremont. But Fred didn't want to leave Dick's until the engine had been tested to see if the repairs had done the trick. After the usual two to three hours to get steam up, the engine was ready for the road. It was a beautiful, clear May morning and every ridge and gulley on Coniston Old Man stood out in sharp relief. But Fred was oblivious to the magnificent Lakeland scenery. He'd always said that to him one mountain looked just like another and this morning he'd got far more important things to concentrate on. Alf and Jimmy hitched up the living van; Fred climbed up on to the engine and pulled it out on to the road that they'd struggled along two days earlier. He got into top gear and drove up and down the road and soon he was smiling; to his great joy and satisfaction the tractor was performing the way he had always hoped it would. The adjustments he'd made to the cylinder had worked and for the first time it felt like an engine that he could take some real pride in.

Jack drove the engine up on to the low-loader and set off in it with his dad for Egremont. But it was late afternoon before they arrived. They'd been driving past the Ravenglass and Eskdale Railway, which had always been a great favourite not just of Fred but of Jack and Roger, so they'd made a short detour. It was a place where Fred had spent many happy days with his lads and it was difficult to drive past for what was likely to be the last time without taking a quick look. By the time we'd done some filming in the mine and some driving shots it was nearly nine o'clock in the evening. The day had been far too long for Fred and it wasn't over yet. He had an appointment at Christie's Hospital in Manchester at nine the following morning. Jimmy was taking Fred back to Bolton that night so that he could take him to the hospital and bring him back up to Cumbria the next day. Fred was worried about his engine; he didn't think the site was secure so Alf volunteered to stay in the living van as nightwatchman.

Next morning it was off to Workington steelworks to see where the raw material that used to be mined at Florence was turned into steel for railway lines. Today most of it is rolled in a modern computer-controlled rolling mill, but what Fred was interested in was seeing the old hand-rolling mill which is still used for small light and narrow-gauge railways. We started filming while Fred was back in Manchester at the hospital. Jimmy drove him up straight after his appointment. We set up by the rolling mill in time for Fred's arrival at the works. It was the one he had enthused about so many times; the one we'd filmed him with at Ironbridge when he did his magnificent action replay of what it had been like when he'd seen it working at Walmsley's in Bolton, its original home. This one at Workington was the only model left in the country that was still in operation. Fred hadn't seen anything like this in action since he was a boy and I wanted to capture his first reactions when he saw it. The camera started to roll as he walked in, but when he saw the rolling mill he was subdued. Two or three years earlier there would have been no stopping Fred from getting the tongs from one of the workers and having a go at handling the white-hot metal himself as it came through the rollers. But not now. He livened up a little bit as he talked to one of the workers, but overall there was a terrible sadness about him.

It was the end of the first week on the road. We'd known it was going to be difficult, but nobody could have anticipated quite how difficult. The engine was performing better after the work at Dick Ransome's but it was still slow. Everything was taking so much longer than expected – getting the engine steamed up, cleaning it down and sheeting it up at the end of the day, loading and unloading it from the low-loader. The logistics of arranging a filming schedule around a traction engine were difficult enough and timings were impossible to keep to, but when this was added to looking after a very sick man who was the central character in the films I wondered if it would be possible to carry on. In all of this Fred's health and well-being had to be our first concern and, although he was bravely trying to hide the fact from everybody, he was clearly in a lot of pain. Fred knew that this was the last time he would see old friends like Dick Ransome and he was unhappy that the demands of the engine and the filming schedule hadn't given him enough time with him and with Roger Mallinson. But he was determined to carry on. There were more old friends to see on this trip and he wanted to visit all the small firms that had helped him with parts or advice for the engine. This was Fred's way of saying thank you.

If we were to go on there would have to be a big rethink about schedules, about the number of places we could visit, about the distances involved and the amount of driving that Fred would be able to do on the engine. The next stage of the journey was to be a long one, up to Bo'ness on the Firth of Forth, where we planned to visit a foundry to see how the castings for an engine like Fred's were made. Before attempting that sort of distance we needed to do some rescheduling and Fred needed some time to rest back at home and do some fine tuning of the engine to get it performing better. So all the filming was put back by two weeks.

The engine was loaded up on to Alan's low-loader and taken back to Fred's, where he was able to do some more work. One of his worries concerned leaks from some of the rivets, so he put various concoctions into the boiler, including porridge oats, red lead and large quantities of urine gathered in a bucket at the back of the shed. These were old recipes he'd found from studying the methods of the traction engine men of the past and, after two weeks' work and experimentation, Fred was convinced things would be better.

When Alan came to pick up the engine to transport it to Scotland, Fred was in good spirits. The rest had been good for him and for the engine. On the way up to Bo'ness, Alan's low-loader stopped at Abingdon so we could get some driving shots of the engine in the Scottish Borders. The good news was that the two weeks' work on the engine had improved performance massively. It bowled along through the hills in fine style. Fred was much happier and there was no talk of selling it now. But there was a problem. Fred, Alf and Alan had been up at the crack of dawn to get the engine onto the low-loader. But by the time they had loaded it up, driven to Scotland, unloaded the engine, got it steamed up and driven it round for the shots we wanted it was nearly six o'clock and they were still around eighty miles from their destination.

I drove on ahead to the Bo'ness and Kinneil Railway, where the engine was to spend the next few nights, to make sure there was somebody to let it in. The railway offered good secure parking and a plentiful supply of coal and water, and workshops if any running repairs were needed. But some of the roads from Abingdon to Bo'ness were narrow, which made the journey for the low-loader slow, and the boys didn't arrive with the engine until nearly nine o'clock. Again it was too late for Fred in his condition. It had been far too long a day, so another rethink would be needed about getting driving shots.

The highlight of the Scottish section of the journey was a visit to the Forth Bridge. Fred would drive the engine over the bridge; it would be the only traction engine ever to have done this under its own steam. He was then going to climb to the top of one of the towers. Fred was looking forward to it; he still got a buzz from being in high places. We'd already been to the top of the Forth Railway Bridge, so this would complete an interesting double. I went to the tower first, to film the shots of the engine crossing the bridge from the top. I went up with a rigger and, just like the Humber Bridge, which we'd done a few years earlier, there was a lift. On the Humber Bridge the lift goes all the way to the top of the towers but, when I stepped out of the lift in the Forth Bridge, we were still inside the tower. The last thirty feet or so was up a vertical ladder attached to the inside of the tower. I looked at it and knew Fred wouldn't be able to do it. It was heartbreaking. Everybody's favourite steeplejack had spent his life climbing three-hundred-foot chimneys and he wasn't going to be able to get up this. In the earlier days of our filming together he would have raced up, but he couldn't do it now, especially after driving the engine all day. It was moments like this throughout that last shoot when it came home to me forcefully how close to the end Fred was. Throughout the shoot he was in denial. He was using the tour to keep himself occupied with something he enjoyed doing and to keep all thoughts of his failing powers out of his mind. I didn't want to bring him up this tower and face him with something he could no longer do.

Fortunately, on this one occasion, the slow progress and poor performance of the engine came to my aid. When Fred and Alf got over to the Fife shore, steam pressure was low and they couldn't attempt the return across the bridge until they had raised it. This delay meant that by the time they got back across the bridge it was too late to go up the tower. Fred was disappointed, but he knew he needed to get back to our hotel for some rest. Once the engine had cooled down it was sheeted up and left in a secure parking place near the bridgemaster's office.

The next stage of the journey was to take Fred to the north-east of England. Along with coal, it's water that's needed to raise the steam that powers an engine on its journey around Britain. So Fred's next port of call was a great pumping station that had been built to cater for the ever-increasing demand for water in the Victorian Age. Then it was on to Bradford to look at one of the most important parts of any steam engine: the boiler. Until quite recently there was a Boilermakers' Union, but the trade has all but disappeared now. So Fred went to see one of the few boilermakers left in Britain – Israel Newton's Boiler Works, where boilers are still made in the traditional way, using rivets rather than welding.

On the way Fred told us about the importance of good boilermaking. 'The backbone of my engine is the boiler which is needed to raise the steam to make the thing go.

Boilers contain steam at a very high pressure so you can be lighting a fire in something that's very dangerous and end up, as a man in America did, killing five people, including himself and his own son.' Fred was still getting irritated by leaks that kept appearing in his own boiler. 'These holes are only as big as flies' legs,' he said. 'They're not great big gaping holes, but they're a bit of a nuisance. But the boiler inspector mustn't think it's going to blow up so we'll be all right.'

Fred's boilermaker friend Alan McEwen remembers Fred ringing him at this time. 'He had one or two tiny leaks emanating from rivets on the traction engine he'd so lovingly put together, and he rang me up, and I heard him shouting down the mobile, because he wasn't used to talking to mobile phones, "Alan, I'm in real trouble." So I said to him, "You've no problem whatsoever; there's no problem with that. Do the following." There was nothing wrong whatsoever with his craftsmanship and his boilermaking expertise.' It was just that Fred was a perfectionist. In the olden days new boilers were put out in the rain to go rusty for twelve months, which sealed them up. But Fred didn't have twelve months.

Traction engines are an expensive business but Fred had never been afraid to have a go himself. Of all the work he did on the engine, he reckoned the boiler was by far the most challenging part. At Israel Newton's Boiler Works boilers are still made in the traditional way, using plate-bending, flanging and hot riveting techniques. Fred was proud of his own boilermaking skills and when he arrived at the boiler works the first thing he wanted to do was to show off all the best bits of his own engine.

From Bradford the next stage of the journey took him on to South Yorkshire. The name of Sheffield is associated with steel all over the world and it was the advances achieved there in steel manufacturing that made the mass production of steam locomotives, winding engines and traction engines possible. So that was to be his next stop. On his way he wanted to call in at the Bolt and Rivet Manufacturing Company in Worsborough, where they made all the rivets he used when he was rebuilding the engine. 'When I rebuilt my boiler,' he said, 'I had a great deal of trouble finding rivets, especially made of the right material. At one time every little town and hamlet had its own nut and bolt maker and rivet maker but the invention of electric arc-welding put all these men out of business. Once rivets were crucial; they held together everything from things like my boiler to the Forth Bridge.'

A couple of days' filming in Sheffield took Fred through to the halfway stage of the tour. From Sheffield he was going down into Derbyshire, where he planned to stay for a few days with his old friend Ian Howard of Alton Engineering and visit the workshops of the Midland Railway Centre. By this time I'd handed some of the directing over to Jon Doyle, the assistant producer on the series. Fred got on well with Jon, as did the rest of the team. Fred liked him because he'd taken the trouble to get to know a bit about how the engine worked and he'd filmed all the preparations for his last chimney-felling. Nobody had ever done that before and Fred was pleased that there was a record of it.

It was the end of the week, Friday night. Fred and the team were staying in Belper. Next day they were due at the Midland Railway Centre, but at about ten o'clock I got a phone call at home. It was Jon. He was very worried about Fred and didn't think he would be able to carry on. His legs were very swollen and his feet were so bad that he couldn't get his boots on. It sounded as though it was to be the end of the journey. I drove to Derbyshire early the next morning, getting to the hotel before breakfast. Alf was first down and he confirmed how bad Fred was. When I saw Fred he agreed that he couldn't continue. It was Saturday 26 June. He was due to receive his MBE on Wednesday 7 July. For Fred this was going to be one of the most important days in his life. The priority was to get him better for that. Forget about any more filming before then, I told him. He needn't worry about the engine. Alan had taken it from Sheffield to the Midland Railway Centre the previous afternoon and it could stay there, securely parked, for as long as we wanted. The low-loader would pick it up the day before the MBE ceremony and transport it to London so that Fred could drive it.

I took Fred and Alf back to Bolton that morning and insisted that Fred went to see a doctor straight away. When I returned to his house early the next week Fred was looking a lot better. He'd been given some tablets for the swelling and they had started to take effect. He'd also got some stronger painkillers. He was confident that he would be all right for the big day. But he also wanted to carry on with his grand tour once he'd had his appointment at the palace.

I wasn't sure about this. At the rate we were currently working, it would involve at least another two months' filming if we were to do it three or four days at a time with breaks in between for Fred to rest. His condition had deteriorated rapidly since we had started the journey less than two months previously and I just didn't think Fred would be able to keep going long enough. But the stubbornness and determination and single-mindedness that had characterized Fred throughout his life came through yet again. Completing this journey had become his last great obsession. There were still too many places he wanted to go to; too many friends he still wanted to see. I talked to Sheila and to some of Fred's closest friends. They all said that it was this journey that was keeping Fred going. Everybody felt we should carry on – nobody more so than Fred himself.

16

The Final Journey

The original schedule for the rest of the journey had involved picking up where we had left off in Derbyshire before going to the Forest of Dean to see some of the earliest methods of smelting the iron used for building the engine, then on across the Brecon Beacons to see Blaenavon with its coal mine and ironworks. Much of what you see on a traction engine is tin, so Cornwall, the great centre of the tin industry, was to be Fred's next stop. Then we were to head back up through the West Midlands and North Wales and home to Bolton. But from now on we realized we would need to stay within a short drive of Bolton so that Fred could be quickly transported back home if necessary. Added to this, sitting in a car for long periods was causing Fred a great deal of discomfort. South Wales and Cornwall were clearly too far and would have to be dropped from the schedule.

The problems involved in looking after the needs of a

sick man while arranging a filming schedule around a traction engine had been enormous. Getting a traction engine around involved long hours and physical effort. It was something that Fred was no longer capable of doing himself. Filming meant working to tight schedules and also putting in long hours, especially when waiting for a traction engine that took a long time to get anywhere. From this stage on, therefore, we would have to operate as three separate units: Fred, the engine and the production team.

Fred's welfare had to be our primary concern so I arranged with Jimmy that he would act as his chauffeur. The instruction was that he didn't need to worry too much about the time that the crew was scheduled to arrive at a place and start filming. Fred needed to be able to do things at his own pace. If he needed a break in the journey he could stop without worrying about the schedule. If he needed to get back to Bolton quickly to see a doctor, Jimmy would be able to drive him straight back.

One of the things that gave Fred great pleasure on this trip was driving the engine. It was now performing well, steaming along with relative ease at the 12 mph he had originally anticipated, and he was enjoying the experience. 'It's quite strange really,' he said, 'because I've never driven a traction engine of my own even though I've driven quite a few dozen belonging to different mates. And of course it's quite nice when you get a chance to drive your own after twenty-odd years of struggling and restorations to get it into the state that I want it in, like a new one as you might say.' But, much as Fred loved being on the engine, his health wouldn't allow him to drive it long distances and he certainly wasn't fit enough to do all the hard work involved in getting it loaded and unloaded from the low-loader and getting it steamed up and then cleaned down at the end of the day. So his engine needed to be transported to all the locations and steamed up ready and waiting for Fred. He'd do a short drive on it and be seen arriving at the location and departing from it. At the end of each day the engine would have to be in secure parking because Fred couldn't stay with it in the living van. The production team would travel round as a separate unit, arriving at locations at the scheduled time and starting to set up and do some of the filming while waiting for Fred.

As Fred made preparations for his MBE ceremony, I worked out new schedules for July and August. We would not be travelling too far and Fred would not be away from home for more than three or four days at a time. The only long trip that was left was the journey down to London for the MBE. He was going to go and collect it on his beloved engine. We wanted the tractor to steam up the Mall and through the gates of the palace, but the officials decided that the possible security risks of driving a potential bomb into the palace were too great. The Royal Parks Commission were also not too keen because of the damage that a twelve-ton steam engine might do to the road surface of the Mall. But the army came to the rescue. The adjutant to the Irish Guards, Alex Turner, offered a parking place for the engine and living van on the parade ground at Wellington Barracks on Birdcage Walk, just a minute's march away from the palace.

One of the great backup services that Alan Atkinson provided when we arrived at a location was to do a brew for everyone. He always kept a portable gas stove, kettle, water, tea and coffee in his cab and when we arrived on the parade ground at Wellington Barracks, Alan put a brew on just as he did everywhere else. Fred got his picnic chair out and, brew in hand, he held court with the commanding officer who had agreed to us being there. Then with the camera ready to roll and the Irish Guards marching past the van practising Changing the Guard, Fred got into the living van in his cap and oily driver's jacket and emerged wearing his morning suit, ready to walk the few hundred yards over to the palace to receive his gong.

As Fred walked up to the gates of Buckingham Palace on Wednesday 7 July 2004 accompanied by his wife, Sheila, and his sons from his second marriage, Jack and Roger, he was a very proud man. He had always been a great royalist and he talked with pride about the days when Britain had an empire, so being made a Member of the British Empire had great significance for him. After the ceremony the newspaper photographers and television journalists queued up for interviews. The interviewer from Channel 4 News wanted to know if it bothered Fred that his medal's title had this empire word in it. 'No, no, it doesn't,' said Fred. 'I'm a royalist, mate, I'll tell you.' When the interviewer persisted and said, 'A lot of people don't take the honour because they don't like the word empire in it,' Fred's pride in empire and royalty shone through. 'Yeah,' he said, 'when we were running it, it all seemed to work a lot better than what it is now.' When he was asked by Sky News what he'd received the award for, he replied simply, 'I got it for back-street mechanicking.'

There in the courtyard of Buckingham Palace I talked to Fred about the filming we'd done together over the years and about the things he had enjoyed most. 'A lot of it's been enjoyable,' he said, 'but the great drawback with television presenting is you don't get enough time to actually look at a place properly or study the objects that are there. When we went to the Science Museum here in London they had to keep coming to find me because I'd wandered away to look at things. I could have spent days there looking at all that stuff, not hours. But of all the things we've been to I think the lantern on Ely Cathedral were quite a magnificent piece of woodwork to me. That's why we created that model that we did. That were my version of how them men got it all up there in the Middle Ages. Then there were the more modern things like the Forth Railway Bridge and even the more modern, modern ones like the suspension bridges made of wire, which are quite fantastic things.'

After the ceremony Fred couldn't wait to get out of his

posh morning suit and back into his cap and working clothes to resume his tour of the country. But first of all he celebrated receiving his MBE by touring the sights of London on the engine with Jack and Roger on board. From Wellington Barracks it was a quick spin down Birdcage Walk to Parliament Square, past Westminster Abbey and over the Thames at Lambeth Bridge, back over Westminster Bridge past the London Eye and Big Ben. As usual it was thirsty work on the engine so, on the way up Northumberland Avenue, Fred spotted a pub and, to the consternation of the traffic wardens, he parked up, leaving Alf to look after the engine. 'Tell them it's a refuelling stop,' he said to Alf, 'and if it gets too thirsty it'll blow up.'

Suitably refreshed, the tour continued past Nelson's Column, down Whitehall and back to Parliament Square for another circuit. For the tourists it was an unexpected bonus and cameras clicked to get shots of the engine chuff-chuffing past the famous sights. For Fred it was just a wonderful outing and for a short time he was able to put his problems out of his mind and enjoy himself. 'The whole family had a great day out and it was a fantastic occasion,' he said as we returned to Wellington Barracks to get the engine back on to Alan Atkinson's low-loader. 'I couldn't think of a better way to celebrate than touring around London showing everyone my traction engine. It was good to see their faces when they saw my engine coming down the street with all the red buses and black cabs.'

The following week we picked up on our filming again. The rest that he'd had and the MBE seemed to have given Fred a real lift and he was given a further boost when Jack and Roger joined him for the rest of the trip. At Great Central Railway, which runs from Loughborough to Leicester, they polished the engine with Alf and Jimmy as Fred had a ride up and down the line on the footplate of a 1912 04 Class locomotive. When he got back he had a can of Guinness and sat in one of his picnic chairs in front of his tractor while he looked at the engine sheds with Roger and told him about the engine sheds in Bolton when he was a lad. He told him about the lamp that he'd got from an old engine driver and promised him that one day it would be his.

The following week Fred and the boys stayed with Ian Howard at Kirk Ireton in Derbyshire. Ian was an old friend of the family. His business, Alton Engineering, specialized in the restoration and repair of traction engines. Jack and Roger had known Ian's lads since they were small and it was very much a home from home for them and for Fred. We'd got filming to do at the workshops of the Midland Railway and several other locations nearby and Ian's house provided a friendly and comfortable base.

Ian and his brother Gary owned a magnificent Showman's Engine that had been destroyed in a fire at their workshops a year earlier. Ever since they'd worked tirelessly on restoring it to its former glory and when we were at their workshops the job was nearing completion.

Fred told them how he had cried when he first heard about the fire and the terrible damage it had caused. For a non-steam man like me, it was a humbling experience to see the dedication these men had. They were men like Fred, men with a magnificent obsession, and it was impossible not to be filled with admiration for all the hard graft they put into preserving these machines they loved.

For Fred having his engine somewhere like this at this stage of the journey was very useful. Although it was steaming well now, other faults had started to appear. He was particularly worried about a problem with one of the bearings. Fred was able to leave the engine in the workshop while we went out filming with him. Jack stayed behind to look at the bearing. He'd learned well from his dad and he was a skilled mechanic, due to start an engineering apprenticeship with the Isle of Man Railway as soon as the filming had been completed. Fred told us what Jack had been doing with the engine. 'Jack took it upon himself to take off the front wheel and have a look at a problem we have of the liquid brass coming out the bearing. We've known about it for a week or two and it's in quite a mess really. It'll not stop us carrying on but there's something terrible happened in between the brass bearing in the hub of the wheel and the actual shaft. It's lost about a thirty-second of an inch underneath, with the weight of the engine resting on it, but when we get back home we're going to have to do some serious mechanicking.'

So the engine would be out of service for a few days

but, while Jack was working on it, Fred had got places he wanted to visit. 'I'm going down the road to see Mr David Ragsdale, who's the owner of six traction engines, and the main reason we're going is he's a master at the art of pattern-making. From ornamental park railings to the cylinder block of a traction engine, everything that was made out of iron had to be cast. The casting was done in a mould and to create the mould a pattern had to be made. Once they'd made the pattern, they could use it dozens of times. The thing is that, in the 1920s and 1930s, when the demise of the traction engine were very obvious to most people, they all got burned, so anybody now who buys an engine that's got a dicky part has got to come here and see Mr Ragsdale, who specializes in making patterns from the remains of the original bit or a drawing.'

Back at base, the wheel was ready to go back on to the engine but Fred wasn't happy that he'd had a problem in the first place. 'It's a bit disappointing, that,' he said, 'but it's one of the simpler bits of engineering that's somehow gone completely haywire. I wouldn't like to do another five thousand miles with it!'

With the wheel back on it was time for the road again. We visited the Severn Valley Railway and the Black Country Living Museum. Fred's engine is steered by chains, 'so we've come to the Black Country Museum,' he said, 'to see a gentleman making chains in exactly the same way as the chain on my engine would have been made in 1912, without any fancy electric welding or anything like that.' This was our fourth time of filming at the museum and it had always been one of Fred's favourite locations. They have a very fine old pub and just over the road there's an excellent fish and chip shop. 'Good ale, fish 'n' chips in a lovely place,' was Fred's verdict. 'I bet half of them men in London with their fancy bloody suits on and their fancy shirts have longed for this. They might make a lot of money but the bloody stress of it all must be terrible.'

Ian Walden is the director of the museum and he talked about the contribution that Fred had made through his many visits. 'The Black Country Living Museum has a particular role to play in the history of the Black Country and the cradle of the Industrial Revolution but telling that story is very difficult. It's so complex and there are so many things involved, from steam to canals to buildings, and Fred helped us no end in putting it over and through his enthusiasm almost convinced us we were doing the right thing. I'm sure he did that with lots of other people. If somebody was a volunteer restoring a steam engine or anything else like that Fred could make you feel you were the most important person in the world for doing it and we owe him a lot.'

It was appropriate that his next port of call was the headquarters of the North Staffs and Cheshire Traction Engine Club, where he would meet up with a lot of his old mates from the traction engine world; all people like him who devoted their lives to keeping our steam heritage alive. They'd all got their engines in steam, many of them working on jobs that needed doing around the site. As I watched Fred with his mates and their engines I knew it had been the right decision to go on with the filming. He'd got his own lads with him and there couldn't have been a better way for him to spend his final days.

Fred had always told me how he would get a barrel of beer whenever he was going to a rally and set it up on the running board of the van. We'd been staying in Burton on Trent the night before, just opposite the Burton Bridge Brewery, so we bought a barrel of their best bitter and set it up on the living van as Fred had done in days of old. Fred got the picnic chairs out and sat, pint in hand, talking to old mates. Alf and Jimmy pulled the pints. It was a hot July afternoon in the heart of the English countryside and Fred was surrounded by old friends and their engines. The film crew was forgotten as old stories were told and old friends remembered. Len Crane, who'd transported Fred's roller so many times in the past and restored the Bratch Pumping Station as well as his own steam crane, was there with his wife Jane. Like Fred, Jane was suffering from cancer. They talked of happy times in the past, but they also talked of the future and of a new generation of enthusiasts like Jack and Roger who were going to keep the steam movement alive.

Fred was proud of his lads and by this time they had become essential members of the team. They would usually travel around in the cab of the low-loader with Alan and Alf so that they could be with the engine all the time, ready to help with loading and unloading it, lighting

the fire and getting steam up when we arrived on a
location and cleaning the engine at the end of each day.
But much more important than that, they were providing
much-needed support for their dad. Fred thought the
world of his lads and it was a great comfort to him that
they were able to be with him on the final stages of his
journey. He'd been separated from them for nearly eight
years, only seeing them during the school holidays and
occasional weekends, so he took real pleasure in going
with them to the places he loved and cared about. He
positively glowed with pride as he watched them driving
and steering the engine and looking after it.

Everywhere we went he would say to Jack and Roger,
'One day my engines will be yours.' The engines were
the only things he referred to. He didn't talk about money
or royalties or even about what was going to happen to
his house and garden. For Fred the only thing that really
counted was those two engines that he had devoted so
much of his life to and the fact that they would go to Jack
and Roger.

Surrounding Fred on the journey was a great support
team. Nobody who was there will forget the dedication of
Alf, Jimmy and Alan. Nothing was too much trouble for
them if it helped their friend to achieve one of his last
ambitions. It was a fine example of the sort of friendship
that Fred valued so much: the comradeship that men like
miners have, the kind of support that comes from depending on your workmates in difficult and dangerous
conditions. They were values that had shone through in

his programmes; values that he thought had been lost as everybody worked on computers in offices, but nowhere were they more in evidence than now.

But there were no more late nights at the pub; no more lock-ins. Fred would still have a few drinks at the end of the day, but it would usually be in the bar of the hotel we were staying in. He'd come down and talk over a few pints and, although he'd still tell anybody who cared to listen about the 'twenty-seven years and two divorces', there weren't as many of his stories now. He was frightened by the terrible things that were happening to his body and he was confused. There were people mithering him, he said, about who he should leave things to; telling him this, that and the other about what he should do with his house and his garden and everything else. Occasionally he would ask for advice but he had to be told that these were decisions that he alone could make. Nobody could make them for him. But generally it was not a subject he wanted to dwell on. Throughout this period, thinking about a will or making sure his affairs were in order was not something he wanted to be reminded of. For Fred, these things meant facing up to the inevitability of imminent death and he wasn't prepared to do this. There were still too many things to be done. Many a time, sitting in the bar at the end of a day's shoot or at his kitchen table between shooting trips, he would tell me about an article he'd read in the paper about another new drug that offered hope to cancer sufferers. Maybe he'd be able to get it in time.

But the reality was that Fred's condition was growing worse by the day. By the end of July we had done thirty-four days' filming with Fred out of a total of sixty that had been scheduled for the series. There was clearly not a chance that he would be able to do the full sixty. I talked to him about it. If we did three more days in North Wales I said I'd have enough for the twelve-part series we had been commissioned to make. Fred was reluctant to bring the journey to an end, but by now he was in more and more pain and it was all getting too difficult. He wanted to spend some time in his garden. He knew it was near the end, but there were still jobs he wanted to do and places he wanted to visit for the series. He definitely wanted to do the North Wales trip because we were planning to steam past Snowdon and over the Pass of Llanberis to the former Dinorwic Slate Quarry, where Fred wanted to show us how the sort of Victorian workshop where his engine could have been built would operate.

As well as North Wales, there were two other places nearer to home that he insisted we should go to. Both of them had played an important part in the building of his engine. Fred had always picked out Bowns Dukinfield Ironworks for special praise. 'Their business is metal pressings,' he said, 'and they made the throat plate and ash pan for my engine. They are a real old-style engineering works that's been in the same family for five generations, the kind of outfit the steam men should support.' Fred wanted to make sure they were included in the series.

Some of the most important parts of the engine are the smallest, like the nuts and bolts that hold it together, and Fred wanted to go to the place where he had got all the nuts and bolts for his engine. Thomas Smith's, or Smith Bullough as it's now known, is one of the old-time nut and bolt works that produce about eight hundred bolts a day. One of the problems that steam enthusiasts have is that it is difficult to get the right nuts and bolts for their engines because everything is metric now. The beauty of Smith Bullough's workshop is that they still make the old-style imperial thread sizes that a traction engine like Fred's needs.

We scheduled two days' filming in North Wales. On the way back to Bolton we would stop at the Anderton Boat Lift in Cheshire, then go to Dukinfield Ironworks the next day. But Fred was still driven by his obsession to get everything done in the garden, so rather than having a rest between filming he decided that his pithead gear needed a coat of creosote. Despite many offers of help he insisted on doing the job himself. He managed to climb the ladder but the combined effects of two gruelling cycles of chemotherapy, rapidly failing strength and the painkilling drugs that he was taking meant that he just wasn't up to doing the sort of thing he'd done with ease throughout his life. As he was up at the top he over-stretched, the ladder slipped and Fred fell about twenty feet, landing on his back near the top of his mineshaft. At first he couldn't move; he'd hurt his back and he was covered from head to toe in creosote. Fortunately Jack

In a local pub later in the day Fred talked to some retired workers from the workshop. One of them told Fred how he hadn't known anything about foundry work, until a foundryman came down from Scotland and showed him all about casting brass. Fred returned to one of his perennial themes for what turned out to be the last time on film. 'Well, that's the best way,' he said. 'We've discussed this a lot on this trip, about learning out of books and things like that. You can get a book written by an academic who's never done anything in his bloody life. I mean OK, he's a clever bloke, he can write it all down, but it doesn't really give you the final bit of how to do it. Having an old man stood at the side of you saying, "You're making a bugger of that," is the best way to learn.'

Fred was now tiring quickly and he was clearly in a lot of pain, wincing visibly as he tried to sit down or stand up. It had to be bad because he had always given the impression that he was completely oblivious to pain. Jack expressed his concern to me. Unlike anybody else, he and Roger were seeing their dad in his room and they could see the extent of the bruising on his back and the swelling of the tumour. I told Fred that he shouldn't be doing this. He should have let me know about the fall. There would be no problem postponing the rest of the filming until the bruising on his back had gone down and he could move around more comfortably. But he insisted on going on.

The next day Fred and his team went to Anglesey to see a mine, where a lot of the copper for a traction engine like his would have come from. At one time Parys Mountain was the largest copper mine in the world. Parys Mine and the nearby Mona Mine dominated the world's markets in the late 1700s. Today the great opencast is an impressive chasm, opened up at an early stage of mining after a collapse of other workings. But there are still many abandoned workings that are being charted by a mine exploration group. Fred had always been fascinated by old mine workings, but he wasn't fit to go underground. So, after a drive to the edge of the great chasm on his tractor and a look at an old engine house, we went to a nearby pub to talk to some members of the mine exploration group and look at their photographs of the underground workings. But for Fred it was too much. Over the last few weeks we'd watched as the terrible illness had taken its toll, and we'd seen his energy fail and his passion and enthusiasm fade. This afternoon in a little pub on Anglesey it seemed to be extinguished. Fred's tiredness and the effects of the high dosages of morphine he was taking to keep himself going were too much. As we set up the camera and lights to film his chat, Fred struggled to keep his eyes open and finally dropped off to sleep. We couldn't go on any longer. It had to be the end of Fred's brave battle to carry on with his grand tour.

We talked about it that evening over a pint or two in the bar of our hotel. He was still reluctant to give in. The journey and the filming had been a sort of therapy for him. If he accepted that he couldn't do any more he would have to accept the inevitability of death not being far away. He still clutched at straws. There was another new drug he'd read about. 'There's so much to do,' he said and then, 'What have I done to deserve this?' He was desperate for somebody to give him some hope, to assure him that it wasn't all over just yet.

We agreed to stop at Anderton Boat Lift on the way back the next day, but the filming at Smith Bullough and Dukinfield Ironworks would have to be put off till some of the bruising had gone from his back and he was feeling a bit better. The next morning I drove Fred from Anglesey along the North Wales coast. It was a journey we'd made many times before. As we approached the Marble Church at Bodelwydden I slowed down as we neared Fred's favourite steeple, knowing he would be seeing it for the last time. He didn't say anything but, as we drove away up the hill towards Queensferry and Chester, Fred started to talk about his garden. 'I don't know why everyone's so bothered about me garden and about who's going to have it all. It's not worth owt really. Just a pile of junk that I've assembled over the years and rescued from the scrap man.'

Fred's garden was one of the reasons why it was so important to him to do the filming at Smith Bullough. Two weeks later the bruising and the pain in his back had gone down sufficiently for us to make the short journey with him from Bolton to Atherton where the firm was based, and there couldn't have been a more appropriate place for him to do his last day's filming. 'As small workshops like this modernized in the 1970s and 1980s,' Fred explained, 'they had to do away with the old technology like the line shafting that once drove all the machines. I remember coming by here about twenty-odd years ago and it was all lying in the yard and the boss said I could have whatever I wanted. So I was able to provide a good home for all that line shafting. Yeah, it's all going round in my garden now, driving eighteen pieces of machinery.'

Fred was now spending all his time in the garden. Alf and Fred's faithful boiler tender and general help, Ian Thompson, was with him every day. By the middle of August Fred was fading rapidly and, as his health deteriorated, more and more people who'd enjoyed his company over the years were coming to the house to spend some time with him. Sheila was finding what she later described as 'the relentless, constant influx of well-meaning people surrounding him in what were to be his final months' a great strain. 'His mind,' she said, 'was being gradually taken over by the effects of morphine, and on frequent occasions other people's wishes became mixed up with his own and this meant things between Fred and me became uncomfortable. "What are we doing falling out, cock, we shouldn't be arguing at a time like this," he used to say.'

The one day's filming for the series that Fred was still desperately keen to do was at Dukinfield. Because of works holidays and our own holidays it couldn't be arranged in August so a date was set for 2 September. Alan Atkinson would take the engine with Alf and I'd drive from Bolton with Fred and Jimmy. I went over to

Bolton the afternoon before to see how Fred was. I'd been on holiday and had not seen him for a couple of weeks. The change was dramatic. Gone was the bluff, cheerful, healthy-looking Fred I used to know. Now he was pale and sombre. Ian had lit the fire in his boiler and Fred was sitting by it with the old coat on that he normally saved for the coldest days in the garden. It was late afternoon and the first leaves of autumn were beginning to flutter down on to the garden and the shed roofs. Alf and Ian left and I sat with Fred as he looked into the fire. He was frightened and confused and the morphine he was taking was causing him to hallucinate. He was seeing things from the past but everything was very mixed up. What he was clear about was that he still wanted to go to the ironworks the next day to show everybody where the throat plate and ash pan for his beloved tractor had been pressed. 'Them men deserve to be seen,' he said. But a lot of the time he wasn't very coherent and he looked so ill that any filming we did with him now would be unusable. But I couldn't tell him we wouldn't do it. Cancelling the next day's filming would have been like a death sentence.

It was getting dark and before I left I said I'd go to the off-licence to fetch him some cans of Guinness. As I went out of the gate I recognized Fred's old friend Michael Webber. He was standing outside the gate, not sure whether he'd be able to get in to see his former mentor. 'You'd better go in and see him,' I said. Michael stayed and chatted with Fred. Apart from a chance meeting when we'd been filming *Age of Steam*, he'd not seen Fred for a long time. The next morning he helped to get the engine steamed up and went with it on the low-loader to Dukinfield. Michael drove the engine round the streets with Fred standing by his side, a reversal of their roles nearly forty years earlier when Michael had had his first ride on the roller as a schoolboy. It was Fred's last trip on the engine. After riding round the centre of Dukinfield we went through the motions of doing a little bit of filming in the works, covering everything in long-shot as Fred stood watching a pressing. It was the sort of work that he had always admired so much and they were the last shots ever taken of Fred.

Places like this had always been particularly important to Fred, as Anthony Coulls, the collections care manager at Locomotion, emphasized. 'Fred showed people that these traditional skills existed. A lot of people don't understand the work that goes into making a vehicle like Fred's and don't realize that most of those skills are still in existence in Britain today. Many people just didn't know that there were forges still operating; they didn't know that rivets were being made or imperial nuts and bolts or the fact that there are traditional boilermakers still operating where steam engines can be repaired in the traditional manner. Fred just showed people what was in that shed on the industrial estate that was making all that racket and he showed people how we came to be where we are. He also showed that engineering was something not to be hidden away but something to be proud of.'

A couple of days after the filming at Dukinfield, Fred was at home when his condition deteriorated and he was taken into Bolton Hospice. But he wanted to be at home in his big brass bed, near to his garden and his engines and his mineshaft, with his mates calling round to see him. After a fortnight he was granted his wish. Friends, family and old acquaintances crowded round. Fred had always liked talking to people and it was no different now. He still enjoyed the company of his old mates. But he was a shadow of his former self, a frail-looking old man.

Sheila found the pressure difficult. 'I coped alone,' she said later, 'until a time came near the end when I couldn't get near the bed for other people in the room; friends, estranged family, cronies, strangers, hangers-on; all unwilling to leave, taking my place in his life. It got to the point where I would visit Fred early in the morning or last thing at night. With people claiming to know Fred extremely well, and just as many casual acquaintances jumping on the bandwagon and claiming to have a close friendship with him, it was difficult. The trouble was if you'd ever met Fred and talked to him, he would treat you as if he had known you all his life. It was one of his gifts, to make people feel special. If you'd met him twenty years ago – then met again years later, suddenly, bingo! You were close mates, with all the kudos that entailed for people. So lots of people claimed to be his best friend.'

Much as Fred wanted to be there, in the end he

couldn't stay at home. He needed the twenty-four-hour care and attention and the comfort that only the hospice could give. He returned to Bolton Hospice and spent his last days peacefully there, still enjoying the odd can of Guinness and even a glass of red wine. He died there on 6 November 2004, aged sixty-six. He'd finally lost his three-year battle with cancer just weeks after completing the filming on his last series. His friend Dick Ransome, who he'd visited on that last tour, said, 'My greatest admiration is the way that Fred has acquitted himself in the face of adversity with his dreadful illness. He carried on with his work, he finished his steam tractor, he's been on with his mineshaft and one thing and another at Bolton and quite honestly I think that's the mark of a hero. He deserves iconic status to do that, if nothing else, because of his sheer guts and I just hope that he meets his hero Isambard Kingdom Brunel when he goes to the great workshop in the sky.'

Married three times, Fred left behind a wife, Sheila, three daughters, Jayne, Lorna and Caroline, from his marriage to Alison and two sons, Jack and Roger, from his marriage to Sue. Ever since they were young boys, Fred had imbued Jack and Roger with his love of steam engineering and like their dad they had already become experts at the workings of traction engines. On the day before the funeral the boys worked on Fred's engines in his garden, polishing them and tinkering with them in preparation for their role in the funeral procession.

On the morning of the funeral the heavens opened, but

that didn't stop thousands turning out in the pouring rain to pay their last respects as Fred's funeral procession made its way through the centre of Bolton. The town was brought to a standstill as thousands lined the streets. Fred's coffin, with his flat cap on top, was towed from his house to the parish church by the engine on which he had so recently completed his tour of Britain. Driving the engine was Fred's old friend Michael Webber, alongside seventeen-year-old Jack. Michael had taken charge of creating the impressively decorated trailer that carried Fred's coffin. He'd mounted Fred's steeplejack ladders on top to bear the coffin. At the head of the parade was the band of the Lancashire Artillery Volunteers immediately followed by Fred's pride and joy, *Betsy*, complete with living van in tow and thirteen-year-old Roger on the footplate with Neil Carney. Traction engines were bound to play a big part in Fred's funeral and other engines accompanied the procession through the rain-soaked streets.

It was fitting that the funeral service was held at Bolton Parish Church because it was work on the church that had started Fred's steeplejacking career and it was a building he had climbed and repaired on many occasions. There were few sad moments in this celebration of the life and work of a remarkable man. Fred's daughter Lorna read the poem 'Farewell My Friends', dedicated to the many friends Fred had made. The Reverend Barry Newth spoke with great warmth, affection and humour about the man he had first met more than thirty years earlier when he had needed somebody to repair the weathervanes on his church. And heads nodded in agreement around the packed church as I described Fred as 'one of the greatest talents on British television, who remained totally untouched by fame or celebrity'.

The rain didn't let up but the crowds stayed outside the church as the service was relayed to them on loudspeakers. After all, for many Boltonians, Fred's death was like the death of a friend they'd known for many years. This was their Fred, someone they were as likely to have met in a local pub or a scrap yard as they were to have seen on the telly. As the funeral procession made its way from the parish church to the cemetery behind his house where he was to be buried, thousands lined the route, standing silently in the pouring rain. As the coffin went by there were some tears but there were smiles as well as people remembered their favourite moments; maybe it was the bow and arrow he'd used to work out the depth of his mineshaft that they were thinking of, or the arch that collapsed as he sat on it, or maybe it was just 'Did yer like that?' and the cheeky grin as he said it. The stores and businesses of Bolton closed as a mark of respect, and as the procession passed the fire station the firefighters lined up, heads bowed, in front of their engines with blue lights flashing in silent tribute.

Fred was truly a man of the people, a celebrity who always remained one of them, untouched by the fame that television had brought to him. He was buried in a family grave at Tonge Moore cemetery – within sight of the sheds in his beloved garden. As he was lowered to his final resting place all the steam engines that had accompanied the funeral procession gave one single long blast on their whistles. It was exactly what Fred had wished for and it seemed to signal the end of an era – a last post for the spirit and the values of a bygone age.

Fred Dibnah's TV Programmes

With Don Haworth

Fred Dibnah, Steeplejack   1979

Fred   1982

1 The World at Your Feet
2 A Vision of Heaven
3 Uncalled-for Distractions
4 Drinking and Climbing
5 An Epic Journey Begun
6 The Epic Journey Resumed
7 Victory – and After

Fred: A Disappearing World   1983

A Year with Fred   1987

1 Monuments of the Dark Ages
2 A Beautiful Pile of Bricks

3 The Price of Steam
4 Under Pursuit
5 Relief, Ecstasy and Magic

A Year with Fred: New Horizons   1991

Life with Fred   1994

1 Spring Again
2 Early Days
3 On the Road
4 The Big Job

The Fred Dibnah Story   1996

1 Beginnings
2 A Sort of Fame
3 Departures
4 Alone
5 A Reformed Character?
6 Approaching Sixty

With David Hall

Fred Dibnah's Industrial Age   1999

1 Wind, Water and Steam
2 Mills and Factories
3 Iron and Steel
4 Mining

5 Railways
6 Ships and Engineering

Fred Dibnah's Magnificent Monuments   2000

1 Forts and Castles
2 Houses and Palaces
3 Places of Worship
4 Places of Work
5 Bridges and Tunnels
6 Pleasure Palaces

Fred Dibnah's Victorian Heroes   2001

Fred Dibnah's Building of Britain   2002

1 Mighty Cathedrals
2 The Art of Castle Building
3 The Age of the Carpenter
4 Scottish Style
5 Building the Canals
6 Victorian Splendour

Fred Dibnah's Age of Steam   2003

1 The Early Pioneers
2 The Transport Revolution
3 Driving the Wheels of Industry
4 Steaming down the Road
5 Steam on the Water

6 Steam in the Modern Age

Dig with Dibnah 2004

A Tribute to Fred Dibnah 2004

Fred Dibnah's Made in Britain 2005

  1 The Passion of a Lifetime
  2 Collecting the Coal
  3 The Source of the Iron
  4 Castings
  5 Water and Boilers
  6 The Road to Steel City
  7 Mechanics and Riveters
  8 Pattern Making
  9 Engines at Work
10 Chains and Copper
11 The Engineering Workshop
12 A Lifetime's Achievement

Fred Dibnah's World of Steam, Steel and Stone
                                         2006

  1 The Industrial Landscape
  2 Back Street Mechanic
  3 The Machines That Changed the World
  4 Men of Iron
  5 The Victorian Gentleman
  6 Preserving Our Past

FRED DIBNAH'S TV PROGRAMMES

Many of Fred's programmes are available on DVD. For details visit www.theviewfromthenorth.co.uk

Illustration Credits

The author and publishers are grateful to the following sources for permission to reproduce illustrations:

The Paul Donoghue Collection for Fred with Graham, during his art school days, with his parents and Graham, doing his National Service in Germany, as the proud father of two sons, at a pub stop with *Betsy*, *Betsy* outside Fred's house, Fred on his motorbike, the view from the top of Bolton Parish Church, Fred's first big break, his first chimney and the end of the Monton chimney.

The *Bolton Evening News* for the Bolton skyline in 1932 and miners at Victoria Colliery.

Bolton Museums, Art Gallery and Aquarium for *A smoggy day in Bolton* and *Desolate Bolton street* (both from Humphrey Spender's *Worktown*, 1937/38) and for *Beautiful Bolton* and Horwich Loco Works.

The Professional Footballers' Association Collection for L. S. Lowry, *Going to the Match*, on loan to The Lowry, Salford Quays, © The Estate of L. S. Lowry.

David Devine for working on *Betsy* and Ken Devine.

Alf Molyneux for the young steeplejack and Fred in his bosun's chair.

ILLUSTRATION CREDITS

Alan McEwen for preparing for a felling, the chimney at Cockermouth and photos of Fred's garden.

The BBC for Fred's drawing of St Paul's and the views of Blackpool Tower.

The View from the North for preparing for a take with David Hall, last chimney, Llanberis Pass, oiling the engine, a thoughtful moment, running repairs, Jack and Jimmy Crooks lend a hand, Jack, Fred, David Hall and Alf Molyneux take a break, Fred chatting with Gordon Newton, Peak District, life on the road and Independent Forgings.

Kevin Hiley for photos of Fred's garden.

The remaining photographs are from the author's own collection.

The part-title illustrations are Fred's drawings of the pithead gear in his garden and the sinking of his mineshaft.